Inside SCIENCE

Renewable Energy Research

Other titles in the *Inside Science* series:

Climate Change Research
Gene Therapy Research
Space Research
Stem Cell Research
Vaccine Research

Renewable Energy Research

Stuart A. Kallen

ReferencePoint
Press®

San Diego, CA

For more information, contact:
ReferencePoint Press, Inc.
PO Box 27779
San Diego, CA 92198
www. ReferencePointPress.com

LIBRARY OF CONGRESS CATALOGING-IN-PUBLICATION DATA

Kallen, Stuart A., 1955–
  Renewable energy research / by Stuart A. Kallen.
    p. cm. — (Inside science series)
  Includes bibliographical references and index.
  ISBN-13: 978-1-60152-129-3 (hardback)
  ISBN-10: 1-60152-129-4 (hardback)
  1. Renewable energy sources—Research. I. Title.
TJ808.6.K35 2011
333.79'4—dc22
                                                              2010018102

Contents

Foreword   6

Important Events in Renewable
Energy Research   8

Introduction
Investing in the Future   10

Chapter One
What Is Renewable Energy?   14

ChapterTwo
Heat and Light   28

ChapterThree
Wind and Water   41

Chapter Four
A New Generation of Biofuels   53

Chapter Five
Hydrogen and Fuel Cells   65

Source Notes   76

Facts About Renewable Energy   79

Related Organizations   82

For Further Research   86

Index   88

Picture Credits   95

About the Author   96

Foreword

I n 2008, when the Yale Project on Climate Change and the George Mason University Center for Climate Change Communication asked Americans, "Do you think that global warming is happening?" 71 percent of those polled—a significant majority—answered "yes." When the poll was repeated in 2010, only 57 percent of respondents said they believed that global warming was happening. Other recent polls have reported a similar shift in public opinion about climate change.

Although respected scientists and scientific organizations worldwide warn that a buildup of greenhouse gases, mainly caused by human activities, is bringing about potentially dangerous and long-term changes in Earth's climate, it appears that doubt is growing among the general public. What happened to bring about this change in attitude over such a short period of time? Climate change skeptics claim that scientists have greatly overstated the degree and the dangers of global warming. Others argue that powerful special interests are minimizing the problem for political gain. Unlike experiments conducted under strictly controlled conditions in a lab or petri dish, scientific theories, facts, and findings on such a critical topic as climate change are often subject to personal, political, and media bias—whether for good or for ill.

At its core, however, scientific research is not about politics or 30-second sound bites. Scientific research is about questions and measurable observations. Science is the process of discovery and the means for developing a better understanding of ourselves and the world around us. Science strives for facts and conclusions unencumbered by bias, distortion, and political sensibilities. Although sometimes the methods and motivations are flawed, science attempts to develop a body of knowledge that can guide decision makers, enhance daily life, and lay a foundation to aid future generations.

The relevance and the implications of scientific research are profound, as members of the National Academy of Sciences point out in the 2009 edition of *On Being a Scientist: A Guide to Responsible Conduct in Research:*

Some scientific results directly affect the health and well-being of individuals, as in the case of clinical trials or toxicological studies. Science also is used by policy makers and voters to make informed decisions on such pressing issues as climate change, stem cell research, and the mitigation of natural hazards. . . . And even when scientific results have no immediate applications—as when research reveals new information about the universe or the fundamental constituents of matter—new knowledge speaks to our sense of wonder and paves the way for future advances.

The *Inside Science* series provides students with a sense of the painstaking work that goes into scientific research—whether its focus is microscopic cells cultured in a lab or planets far beyond the solar system. Each book in the series examines how scientists work and where that work leads them. Sometimes, the results are positive. Such was the case for Edwin McClure, a once-active high school senior diagnosed with multiple sclerosis, a degenerative disease that leads to difficulties with coordination, speech, and mobility. Thanks to stem cell therapy, in 2009 a healthier McClure strode across a stage to accept his diploma from Virginia Commonwealth University. In some cases, cutting-edge experimental treatments fail with tragic results. This is what occurred in 1999 when 18-year-old Jesse Gelsinger, born with a rare liver disease, died four days after undergoing a newly developed gene therapy technique. Such failures may temporarily halt research, as happened in the Gelsinger case, to allow for investigation and revision. In this and other instances, however, research resumes, often with renewed determination to find answers and solve problems.

Through clear and vivid narrative, carefully selected anecdotes, and direct quotations each book in the *Inside Science* series reinforces the role of scientific research in advancing knowledge and creating a better world. By developing an understanding of science, the responsibilities of the scientist, and how scientific research affects society, today's students will be better prepared for the critical challenges that await them. As members of the National Academy of Sciences state: "The values on which science is based—including honesty, fairness, collegiality, and openness—serve as guides to action in everyday life as well as in research. These values have helped produce a scientific enterprise of unparalleled usefulness, productivity, and creativity. So long as these values are honored, science—and the society it serves—will prosper."

Important Events in Renewable Energy Research

1831
British scientist Michael Faraday uses magnets and a loop of copper wire to produce electric current.

1999
The California Fuel Cell Partnership is created to test and promote fuel cell vehicles.

1979
The Natural Energy Laboratory of Hawaii successfully produces electricity from ocean thermal energy conversion.

1850　　　1900　　　1950　　　2000

1911
The world's first commercial geothermal electric plant is opened in Larderello, Italy.

1954
Bell Laboratories creates the first photovoltaic, or solar, cell to convert sunlight to electricity.

1969
Hydrogen-powered fuel cells provide heat, light, and water for the command module carrying astronauts to the moon.

1987
Geobacter, a microorganism that generates electricity when eating waste products, is discovered.

IMPORTANT EVENTS

2001
Algae is used for the first time to generate large quantities of hydrogen on a continuous basis.

2006
The United States becomes the world's biggest producer of ethanol.

2009
On November 8, high winds blowing across Spain generate 53 percent of the country's electricity for over five hours.

2003
President George W. Bush implements the Hydrogen Fuel Initiative with $1.2 billion for grants to national laboratories, college research facilities, and private companies involved in hydrogen development.

2008
A Virgin Atlantic Boeing 747 flies from London to Amsterdam using a mix of biofuels processed from coconuts and the babassu plant.

2001   **2003**   **2005**   **2007**   **2009**

2002
Engineers at Sky WindPower in San Diego begin building flying electric generators to harvest wind energy from the jet stream.

2007
Engineers drilling for an Enhanced Geothermal System power plant in Basel, Switzerland, cause three earthquakes.

2005
Honda opens the Home Energy Station in Torrance, California, to produce hydrogen for its Clarity FCX.

2010
Chevron Corporation converts an old oil refinery near Bakersfield, California, into an experimental eight-acre solar energy farm.

Investing in the Future

In 2009 the United Nations Environment Programme released a report called *Global Trends in Sustainable Energy Investment*. The report stated that in the previous year, over $155 billion was invested in renewable energy worldwide. According to the report, about $13.5 billion went to companies developing new green energy technologies, while $117 billion was invested in building clean energy projects. Even though the world economy was experiencing one of the worst recessions in generations, the investment in renewable energy was a 400 percent increase over 2004. With billions of dollars flowing to wind, solar, small-hydropower, biomass, fuel cells, and geothermal research, the report left little doubt that renewable energy remained at the forefront of a modern industrial revolution.

For millennia humans have derived countless benefits from renewable energy—power generated by wind, water, sun, and plants, or biomass. But the science of making those elements power modern society is evolving and improving by the day. The next generation of renewable energy is expected to drive the world economy for another century—or even longer. And the competition to develop new energy sources is intense in the United States, China, India, the European Union, and elsewhere.

Higher Oil Prices

Despite the interest in renewables, technologies developed in the nineteenth and twentieth centuries continue to dominate the world energy picture. The infrastructure of modern industrial society was created over the past century to function with cheap and plentiful coal, oil, and natural gas. Worldwide, 95 percent of all vehicles run on petroleum. Coal and other fossil fuels produce three-quarters of all electricity. But fossil fuels are problematic, and some scientists peering into the future see an urgent need to replace these nonrenewable, polluting energy sources with clean, renewable power.

One of the key issues driving renewable energy research concerns the nonrenewable supply of petroleum. In 2010 experts predicted that the worldwide demand for petroleum was expected to increase 2 to 3 percent annually until 2030. In China alone, oil demand in February 2010 was 16 percent higher than in the previous year. However, according to 2008 figures from the U.S. Energy Information Administration, worldwide oil

Renewable energy research and development has attracted billions of dollars worldwide. Manufacturers of wind turbines and solar panels, pictured, are among the beneficiaries of this growing interest in renewable energy.

production is expected to decline 3.7 percent to 6.7 percent a year. Some analysts believe that the decline in production coupled with an increase in demand will cause prices to skyrocket. Americans could face a future where gas costs $10, $20, or even $30 a gallon. According to Philip Dilley, the chair of the oil consulting firm Arup, "We must plan for a world in which oil prices are likely to be both higher and more volatile and where oil prices have the potential to destabilize economic, political and social activity."[1]

The Coal Picture

Although oil supplies are dwindling, coal is abundant. An estimated 1 trillion tons (0.9 trillion metric tons) of the hard, black rock is buried in the earth. Every day across the globe, 14 million tons (12.7 million metric tons) of coal are burned, mainly to generate electricity. In China 2 out of 3 power plants use coal, while in the United States, half of all power plants are coal fired. However, there are many ecological problems associated with coal.

Burning coal is a leading cause of acid rain, global warming, smog, and other air pollution. In an average year a typical coal plant in the United States generates 3.7 million tons (3.36 million metric tons) of carbon dioxide (CO_2), the primary cause of human-made global warming. Each coal plant also produces 10,000 tons (9,071 metric tons) of sulfur dioxide (SO_2). This chemical, responsible for acid rain, damages forests, lakes, and buildings, and it forms small, airborne particles that can harm the lungs. Coal mining is also extremely harmful to the environment. In Appalachia hundreds of square miles of forests and rivers have been destroyed by strip mining, also called mountaintop removal. This process involves blowing up mountains with explosives, removing the coal, and pushing the remaining rubble into river valleys.

A Choice to Make

Because of the problems associated with fossil fuels, even major oil corporations are working to develop a new generation of renewable energy resources. In 2010 Chevron Corporation, the second-largest oil company in the world, converted an 8-acre (3.24ha) site near Bakersfield, California, into an experimental solar energy farm. The land that once contained a polluting oil refinery is now home to 7,700 solar cells of various makes and sizes. They are being tested to determine which ones are most reliable and efficient at producing electricity. When asked about the project, Des King, president of Chevron's technology ventures, stated, "We're quite a large company that uses quite a lot of energy."[2] Although Chevron is one of the world's 6 "supermajor" oil companies, it has been using solar power and hydrogen fuel cells to cut its energy use. And in doing so, the company has saved $3 billion in the past decade.

Chevron is not alone. Oil giants BP and Royal Dutch Shell are also investing billions to develop green energy. And in 2009 the U.S. government increased its commitment to green energy when President Barack Obama signed the American Recovery and Reinvestment Act. The bill included

sustainable energy

Energy produced from renewable sources, such as the sun and wind, through methods that can be used in the future, or sustained over time, without causing environmental harm.

more than $80 billion in clean energy investments, prompting Obama to comment: "[We] have a choice to make. We can remain one of the world's leading importers of foreign oil, or we can make the investments that would allow us to become the world's leading exporter of renewable energy. We can let climate change continue to go unchecked, or we can help stop it."[3] With this bill the United States joined other nations investing billions in renewable energy. In 2009 Germany invested $20 billion, while China committed $440 billion to $660 billion over 10 years. With prosperous nations and major corporations dedicating billions of dollars to renewable energy, the coming decades will undoubtedly see an unprecedented green energy revolution. And the ultimate benefactors will be people, plants, animals, and the environment of planet Earth.

What Is Renewable Energy?

n 2010 computer giant IBM opened its Energy and Utilities Solutions Lab in Beijing, China. The $400 million facility was constructed to develop software to meld China's growing number of solar and wind farms into a "smart" electrical grid. Electrical grids built in the last century connect large, centralized power stations. But the smart grid is a series of high-voltage lines and energy substations that link traditional energy generators with thousands of wind turbines, solar collectors, and other renewable energy sources. This smart grid will be used to increase the percentage of renewable energy powering China's manufacturing sector well into the twenty-first century.

With the opening of the IBM lab, the world's oldest and largest computer firm began an energy partnership with the most rapidly expanding economy on earth. And the enterprise demonstrates China's commitment to renewable energy and CO_2 reduction. IBM vice president Brad Gammons commented on the motivation behind the joint venture: "The Chinese government . . . [wants] to be the market leader in the smart grid."[4]

Power from the Wind

China is the world's third-largest producer of wind power, after the United States and Germany. And with plans to plug the nation's 80 wind farms into a computerized smart grid, China

thermodynamics

The study of the relationship between heat and mechanical energy, and the conversion of one into the other.

is building one of the most advanced renewable energy systems in the world. But scientists studying ways to create high-tech wind turbines are trained in the principles of thermodynamics that were developed in the seventeenth century.

Thermodynamics is the study of the relationship between energy conversion and mechanical actions. It is demonstrated by the way a wind turbine works: The energy of moving wind, called kinetic energy,

A wind turbine spins in the background as a Chinese farmer plows his field. Although China is the world's third-largest producer of wind power, the country's leaders have vowed to dramatically increase wind power capacity in coming years.

is turned into the mechanical energy of rotating blades. The mechanical energy of the blades is converted into electrical energy by a generator located on top of the wind turbine. The generator is based on the principle of electromagnetic induction, discovered by British scientist Michael Faraday in 1831.

Faraday learned that if he rotated a magnet within a loop of copper wire, electric current flowed through the wire. He also found the reverse was true. Rotating wire within a circle of magnets produced the same effect. On a wind turbine, the mechanical energy of rotating blades turns a shaft on a generator. The shaft spins wire inside a magnetic field, which creates electric energy. The resulting electricity is fed into the power grid.

A More Efficient Turbine

While wind turbines play an important role in the total energy picture, the amount of power they produce is limited. This is because it is not

Comparing the Costs

In 2009 the United States got about 49 percent of its electricity from coal, 21 percent from natural gas, 20 percent from nuclear power, and 1 percent from petroleum. Six percent of America's electricity was produced by hydroelectric dams, whereas only 3 percent came from renewable energy. Whatever the source, the price of electricity is measured in cost per kilowatt-hour. A kilowatt-hour is a unit of energy equal to 1,000 watt-hours. One kilowatt-hour hour of electricity would power a 1,000-watt room heater for 60 minutes.

In December 2009 the average price for electricity in the United States was 9.4 cents per kilowatt-hour. According to figures compiled by the *New York Times*, a modern coal plant produces electricity for 7.8 cents per kilowatt-hour; a natural gas plant, 10.6 cents; and a nuclear reactor, 10.8 cents. A wind plant costs 9.9 cents per kilowatt-hour. However, wind farms rely on conventional power plants when the wind is not blowing. This means traditional power plants must be available on hot, windless days when people are running their air conditioners. As a result, the adjusted price of wind energy is over 12 cents per kilowatt-hour, making it more than 50 percent more expensive than coal. Solar remains the most expensive form of energy, costing over 20 cents per kilowatt-hour. However, proponents of renewable energy believe that as technological advances are made, the price of green energy systems will fall and become competitive with fossil fuels.

possible to convert 100 percent of the kinetic energy in the wind into mechanical energy. In theory the most efficient turbine would turn 59 percent of the wind blowing through the blades into mechanical rotating power. This was determined in 1919 by a German physicist Albert Betz, who used his knowledge of physics and math to arrive at that figure.

Despite Betz's theory, no one has been able to build a turbine that is 59 percent efficient in reality. This is because efficiency is sacrificed for cost and reliability. For example, extremely efficient blades, those that rotated very fast, would be very long and light. However, such blades would break in strong winds, making them useless. Because of such real-world conditions, the best wind turbines are 40 percent efficient. And this figure is based on a wind turbine generating power in a relatively strong wind of at least 20 miles per hour (32kph). Weaker wind speeds mean the blades are rotating slower, bringing efficiency down to around 25 percent. This is one of the main problems with wind turbines, as energy researcher Travis Bradford writes: "Large-scale use of wind to generate electricity is, at the moment, limited by the nature of the wind resource itself. Wind is intermittent, which causes the electricity that the wind turbines provide to fluctuate, sometimes dramatically and unpredictably. . . . As a result, wind turbines alone cannot be large-scale providers of electricity."[5]

Energy from Light

Like energy generated from wind, solar power has great potential and several problems. Every day, the sun sends 35,000 times more energy to earth than all human power consumption combined. If utilized, the energy provided in a single 24-hour period would provide enough power to fuel industrial civilization for 27 years. However, in 2010 solar energy only provided about 0.1 percent of the world's daily energy needs. And turning the sun's nearly infinite clean energy into efficient electrical power is one of the major scientific challenges of the twenty-first century.

The solar energy in sunlight can be used to generate two forms of power: heat and electricity. The heat-giving properties of the sun are obvious. But it took the genius of physicist Albert Einstein to devise a theory for converting sunlight into electricity using solar cells. Einstein's 1905 theory of photovoltaic light describes the physical process in which photons, or units of light, are absorbed into materials.

Photo means "light," and *voltaic* means "electricity." In a photovoltaic cell, or solar panel, electricity is generated on an atomic level. Energy from light photons causes the material in the solar cell to release electrons. These free electrons are captured, and an electrical current is generated.

> **photovoltaic**
>
> The conversion of sunlight into electricity.

Einstein won a Nobel Prize in Physics in 1921 for his theory of photovoltaic light. This theory was used in 1954 by Bell Laboratories to create the first photovoltaic cell. Today all solar electricity systems, large and small, are based on the photovoltaic effect. Solar cells create electricity when sunlight hits semiconductors inside the panel. Semiconductors, devices that conduct electricity, are made from silicon, or highly purified sand. All modern electronic devices, including computers, MP3 players, televisions, and cell phones, rely on microchips made from semiconductors. In solar panels the semiconductors are manufactured in such a way as to generate an excess of electrons to create an electrical flow.

The computer industry uses tiny semiconductors in most products. For example, the microprocessor that governs all functions of an Apple

Tiny, lightweight semiconductors, similar to the ones shown here, power computers, cell phones, and other electronic devices. Solar panels also rely on semiconductors, although they are much larger and more costly.

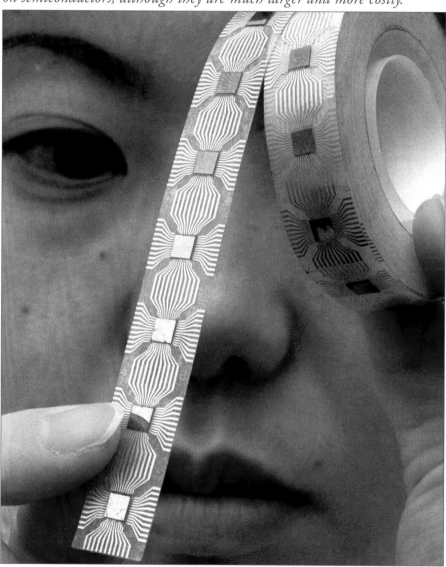

iPod Classic is smaller than a dime. However, in solar panels, the entire surface is covered with semiconductors. Giant semiconductors are very expensive, as is the equipment needed to convert their output into household electricity. And solar cells are not very efficient. Depending on the cell's location, only about 8 to 16 percent of the sunlight that hits a solar panel is converted to electric power. Researchers hope to boost that number someday to 20 or even 30 percent, which is the maximum efficiency possible from solar cells.

Because of manufacturing costs, electricity from solar cells is more than four times as expensive as power from coal. Solar energy also costs twice as much as wind power. In addition, solar cells do not work at night, and their efficiency is greatly diminished on cloudy days.

Water Power

While harnessing the power of the sun may be difficult, people have been using the energy of moving water for millennia. In past centuries flowing rivers powered mill wheels that crushed grain and drove the gears of heavy machinery in factories. In the twentieth century massive hydroelectric dams, such as Hoover Dam on the Arizona-Nevada border, were constructed throughout the world. Today these dams, which use the kinetic energy of water to rotate giant turbines, provide about 18 percent of the world's total electrical power.

> ### kinetic
> Energy that an object has as a result of its motion.

Although dams do not produce pollution, they destroy ecosystems where they are constructed. For example, in 2008 China completed the Three Gorges Dam, the largest hydroelectric power station in the world. While producing as much electricity as 45 average coal burning plants, the lake created behind the dam flooded millions of acres of forest and destroyed the habitat of several endangered species.

Because of the problems associated with large-scale projects, there are few large hydroelectric dams being built in the rivers of the world. However, the ocean is also a great source of kinetic energy, and scientists are hoping to harness the powerful, rhythmic movement of the waves. Ocean power can be used in two ways. Devices can transform the up-and-down motion of waves into electricity. The ebb and flow of the tides also can be captured to spin electric turbines with machines called tidal barrage generators.

Tides and Waves

Tidal power is generally utilized through the construction of gates or small dams built across inlets or reservoirs. The inlets fill when the tide comes in, and as the tide goes out, or ebbs, water is released in a controlled manner. Like a hydroelectric dam, an electric turbine is turned by the water flowing through the system.

There are many benefits to tidal power generation. Little on earth is as predictable as the tides, which rise and fall twice daily on a precise schedule. And with oceans covering 70 percent of earth, the tides create a huge amount of energy that might be harvested. However, tidal barrage systems are very expensive to build, and marine creatures die when they become trapped in the rotating turbines.

Wave power generators are less dependable than tidal barrage systems because waves are not as constant as the tides. However, one promising invention is called an oscillating water column. This uses a propeller device called a Wells turbine. The oscillating water column is a cylindrical shaft placed at an angle in the waves. The turbine is placed on top of the shaft, above the water. When waves flow into the shaft, they push air up the tube, which turns propellers in the turbine. When waves flow out of the shaft, the pressurized air pulls in the opposite direction but continues turning the propellers in the turbine. Both actions generate electricity.

There are about 100 small companies around the world working to develop ocean power. Most are in Europe, where governments subsidize the industry. However, very little electricity is being generated from the ocean, because the systems are unreliable. When there are no waves, there is no power, and big storms can destroy the equipment. But Michael Ottaviano, managing director of Carnegie Wave Energy of Australia, points out, "Wave [power] is probably the last, globally distributed, abundant renewable resource that remains untapped."[6]

Geothermal Energy

Like the tides, the internal heat of the earth's core, or geothermal energy, is constant and abundant. This source has been tapped by geothermal energy plants, which produce about one-quarter of 1 percent of the world's electricity. Despite this low number, a 2007 report by the U.S. Department of Energy (DOE) stated geothermal power could produce as much as 60,000 times the nation's annual energy usage. In 2009 Vice President Joe Biden said geothermal power could play an important part in the

Massive hydroelectric dams such as Hoover Dam (pictured in this aerial view) supply about 18 percent of the world's electricity needs. Each of the 17 generators at Hoover Dam can supply electricity to 100,000 households.

new "clean energy economy that will create a new generation of jobs, reduce dependence on oil and enhance national security."[7]

Geothermal power plants rely on heat that radiates from the center of the earth as radioactive isotopes, including uranium, potassium, and thorium, decay. This heat is responsible for volcanic activity, geysers, and steam vents called fumaroles. In 1904 Italian businessman Piero Ginori Conti used steam emanating from a fumarole to run a geothermal power generator. The device lit 4 lightbulbs in the central Italian town of Larderello. The world's first commercial geothermal electric plant was built in Larderello 7 years later. This remained the only such facility in the world until a geothermal plant was built in New Zealand in 1958. Since that time, geothermal electric generating plants have been built in 24 countries. Five nations—El Salvador, Kenya, the Philippines, Iceland, and Costa Rica—generate more than 15 percent of their electricity from geothermal sources. The world's largest geothermal plant is called the Geysers, a 30-square-mile (78 sq. km) development in Sonoma and Lake counties in the Mayacamas Mountains of northern California.

radioactive isotopes

Atoms that have an unstable nucleus that undergoes radioactive decay and emits heat in the process.

The Geysers is a dry steam plant. It uses pipes sunk deep in the earth that tap into steam that is superheated to about 360°F (182°C). The pressurized steam rotates a turbine that generates electricity for about 1.1 million people. In order to replenish the water taken from beneath the earth, the Geysers complex injects treated sewage water from the city of Santa Rosa and Lake County. This water was formerly discharged into local rivers.

Geothermal power is practically limitless and nonpolluting. And according to a study financed by Google, geothermal has the potential to supply about 15 percent of the nation's electricity by 2030. However, like all other energy sources, there are drawbacks. At the Geysers, when water is injected back into the ground, a process called deep well injection, it creates minor earthquakes. People living around the Geysers experience 20 to 30 earthquakes a year on the magnitude of 3 on the Richter scale. These earthquakes can be felt but do not cause damage. However, in 1973 a much stronger, 4.6 earthquake occurred at the Geysers. Earthquakes of this magnitude cause notable shaking and the rattling of household items.

While the people who live around the Geysers are accustomed to the quakes, new geothermal projects face stiff opposition from local residents. Few people want to live around an electric power plant that causes their walls to shake. And some fear large earthquakes might result if geothermal projects are attempted on a massive scale.

Biomass and Biofuels

Most renewable energy sources, whether geothermal, wind, solar, or hydro, generate electricity. And biomass, a term used for organic matter from plants and animals, can also be used to produce electricity in bioreactor landfills. Bioreactor landfills are built near garbage dumps and capture the landfill gas produced by rotting garbage. Landfill gas, which is about 50 percent methane, is burned to produce electricity in facilities similar to those that use natural gas. An average landfill creates enough methane to power about 10,000 homes.

Bioreactor landfills remain experimental, and only a few dozen pilot projects operate around the world. However, biofuels are used every

Gas from Garbage

Each day the average American throws away about 4.5 pounds (2kg) of trash. Most of this garbage goes to landfills, where organic substances such as food waste, grass clippings, animal waste, paper, and wood products decompose over weeks, months, and years. This rotting garbage produces methane, a toxic gas that contributes to global warming. But methane is the same energy-rich gas that is in natural gas, a fuel that provides about 20 percent of the world's electricity.

To capture landfill gas (also called biogas), piles of garbage are injected with water, which accelerates decomposition. As the rotting garbage releases methane, the gas is captured by an extensive network of pipes and sent to a nearby burner, or bioreactor, where it is used to produce electricity. Bioreactors produce less than 1 percent of all electricity worldwide. However, there were almost 400 operating landfill gas–energy projects in the United States in 2010. And hundreds of new landfill gas–energy projects were in the planning stages in the United States, Europe, China, India, and elsewhere.

day by millions of motorists. The most common biofuel is ethanol, also known as ethyl alcohol. Ethanol is the same substance found in alcoholic beverages and can be made from corn, grains, soybeans, peanuts, sugar-cane, and other plant matter. Ethanol is mixed with gasoline and used as fuel in at least 30 countries throughout the world, including India, Sweden, France, Canada, China, Colombia, and Brazil. Since 2006 the United States has been the world's largest producer of fuel ethanol. In 2009 Americans consumed about 9 billion gallons (34 billion L) of ethanol, mostly produced from corn.

Ethanol is a controversial fuel. In the United States corn used in ethanol production is also used for animal feed. In 2007 demand for corn ethanol pushed corn prices up 70 percent, creating a drastic increase in the price of meat, eggs, and dairy products. In Mexico, which imports most of its grain from the United States, the price of corn tortillas doubled, triggering widespread protests across the nation. Critics also point out that ethanol contains only about two-thirds as much energy per gallon as gasoline. This means cars running on ethanol get about 33 percent lower mileage per gallon than those running on regular gasoline. And

Iceland is one of five nations that generates more than 15 percent of its electricity from geothermal sources. A geothermal power plant provides a fitting backdrop to a warm-water swimming lagoon in Iceland.

manufacturing ethanol is an extremely inefficient process and requires large amounts of oil, natural gas, fertilizers, and other fossil fuels to produce and transport crops. However, proponents of ethanol claim that the process yields 25 percent more energy than it consumes.

Biodiesel

Plant matter can also be used to make diesel fuel called biodiesel, or B100. Proponents of this organic fuel point out that the original diesel engine, invented by German engineer Rudolph Diesel in 1893, ran on peanut oil. Modern diesel engines can also run on peanut, canola, soybean, hemp, and palm oil or animal-based oils such as tallow, lard, and fish oil. While most cars in North America do not have diesel engines, biodiesel can be used in millions of heavy trucks and buses currently on the road. And in Europe, where about half of all cars have diesel engines, millions of cars could use biodiesel as their source of fuel.

Biodiesel is created through an industrial process that involves treating the oil to remove the thick, sticky substance called glycerin. However, in the United States, an estimated 5,000 people run their diesel cars and trucks on waste vegetable oil. This substance is simply old grease from deep fryers used to cook food. Every year, at least 3 billion gallons (11.36 billion L) of waste vegetable oil is produced in the United States at potato processing plants, snack food factories, fast food restaurants, and other eateries. If all that waste vegetable oil could be collected and used to replace petroleum products, it would be equivalent to 1 percent of U.S. oil consumption. While that might be a small amount, proponents of waste vegetable oil, who fill their tanks for free, believe that fry grease can play an important role in the future of renewable energy.

Hydrogen Fuel Cells

Biomass holds great promise because of the abundant supply of organic matter that can be converted to energy. But nothing in the universe is as abundant as hydrogen. About 75 percent of all matter on earth is composed of hydrogen molecules. And hydrogen, which can be derived from the molecules in water, is the main source of power for fuel cells. These high-tech batteries convert the chemical energy of hydrogen into electricity. Fuel cells are nonpolluting; the only by-products from the process are heat and distilled water. Fuel cells can be used to power vehicles or produce electricity for homes, offices, and businesses.

Fuel cell research received a boost in 2003 when President George W. Bush implemented the Hydrogen Fuel Initiative. This program spent $1.2 billion over 5 years to fund government agencies and private businesses working to develop hydrogen production and fuel cell technologies. In 2008 the National Academy of Sciences wrote about the importance of the Hydrogen Fuel Initiative: "A transition to hydrogen as a major fuel in the next 50 years could fundamentally transform the U.S. energy system . . . while reducing environmental impacts, including [global warming gas] emissions, and pollutants."[8]

Despite the optimism of the National Academy of Sciences, there are many problems associated with hydrogen fuel cells. Hydrogen does not exist by itself in nature; it is always found in combination with other elements. For example, water (H_2O) consists of 2 hydrogen molecules (H_2) and 1 oxygen molecule (O). Because hydrogen does not stand alone, hydrogen molecules must be extracted, or cracked, from other substances. In the United States 90 percent of the commercially produced hydrogen is cracked from natural gas, a nonrenewable, polluting fossil fuel.

Hydrogen can also be removed from water through an electrical process called electrolysis. During this process a machine called an electrolyzer uses direct current electricity to split the hydrogen from the oxygen molecule. But this is an energy-intensive process that requires about four times more energy than it produces. And since most electricity is produced by fossil fuels or nuclear energy, electrolysis is a poor method for producing hydrogen to run fuel cells.

> **electrolysis**
>
> A method of separating chemicals, such as oxygen and hydrogen in water, using electricity.

Electrolysis can be performed with renewable energy sources such as wind, solar, geothermal, or biomass. However, there are not enough windmills or solar panels available to produce the amount of hydrogen that would be required to power the nearly 1 billion vehicles expected to be on the road worldwide by 2025. So while nonpolluting fuel cells are attractive because of their ability to run on water, they are less efficient than the lithium batteries currently used in hybrid and electric vehicles.

Years of Research Lie Ahead

In some ways the renewable energy situation in the twenty-first century resembles the dawn of the automobile age in the early 1900s. At that time, people could purchase electric or steam-powered cars, diesels that ran on peanut oil, and cars that were fueled by gasoline. Today solar, wind, geothermal, wave power, and fuel cells are all being used to generate power. Whether or not one system will come to dominate the renewable energy picture the way gasoline did 100 years ago remains to be seen. Whatever the case, replacing existing energy technology with new systems that are affordable and efficient will require years of research and development. But with modern science leading the way, there is little doubt that renewables will play a major role in power production by the end of this century.

Heat and Light

The sun that shines down on the earth and the heat that radiates from the planet's core are virtually infinite renewable energy sources. They are utilized by solar cells and geothermal power plants. But solar and geothermal energy only produce four-tenths of 1 percent of the world's electricity needs. In order to increase that percentage drastically, private companies and governments across the globe are investing billions of dollars in cutting-edge research. Scientists and engineers are working to develop technology that will greatly expand the number of sites where geothermal power plants can be built. In the solar energy sector, researchers are pursuing much higher efficiency in solar energy generation.

Puna Geothermal

In the United States, Hawaii is a perfect test lab for geothermal energy research. The Hawaiian Islands lie above a geological "hot spot" in the earth's mantle that has been volcanically active for 70 million years. The volcanic activity deep beneath the surface of the earth provides unlimited heat for geothermal energy projects. One facility that utilizes this resource is the Puna Geothermal Venture, located on the island of Hawaii, or the Big Island. The Puna Geothermal Venture produces about 30 megawatts of power, about 20 percent of the island's needs, by tapping into underground heat. That is enough electricity for 30,000 residents.

The Puna facility is different from geothermal steam plants located elsewhere in the world. Traditional facilities such as the Geysers in California utilize superheated water with temperatures above 360°F (182°C). The Puna plant uses water at a lower temperature, between 100°F and 300°F (38°C and 149°C). This is an important distinction because lower temperature geothermal resources are much more common than the extremely hot water beneath the Geysers. Researchers think geothermal has great potential beyond Hawaii. By perfecting methods for utilizing low-heat resources, scientists believe they can vastly increase the number of geothermal energy plants throughout the world. That is why binary cycle power plants are at the center of geothermal research.

Binary cycle power plants utilize a device called a heat exchanger. A simple heat exchanger is one small pipe enclosed inside a larger pipe. The large pipe circulates hot water pumped from geothermal sources underground. The smaller pipe is filled with a liquid, called a working fluid, that boils at a low temperature. For example, butane gas, used in cigarette lighters, boils at 33°F (0.56°C). When a working fluid like butane is exposed to the hot water in the heat exchanger, it produces a powerful vapor, like steam. This is directed through a turbine that generates electricity.

Investing in Geothermal

Binary cycle power plants are part of an extensive $350 million research program initiated by the DOE in 2009. Energy Secretary Steven Chu commented on the program:

> The United States is blessed with vast geothermal energy resources, which hold enormous potential to heat our homes and power our economy. These investments in America's technological innovation will allow us to capture more of this clean, carbon free energy at a lower cost than ever before. We will create thousands of jobs, boost our economy and help to jumpstart the geothermal industry across the United States.[9]

In an effort to find new sites for binary and other geothermal power plants, Chu tapped into the resources of the U.S. Geological Survey (USGS). The USGS is perhaps best known for creating detailed topographical maps of the United States. But the scientific agency is also charged with studying the nation's geology, or composition of the earth, and hydrology, the science of water's movement, distribution, and quality. Geology, geography, and hydrology play an important role in developing new geothermal energy technologies. In 2010 the USGS instituted a $30 million program to analyze and classify possible sites for low-temperature geothermal energy plants. The information will be made available to scientists, researchers, and business interests in a nationwide data system.

An Unusual Pairing

Beyond the USGS program, the DOE is spending $140 million to fund 123 geothermal test projects in 39 states. Some of these grants are going

Energy Secretary Steven Chu

Steven Chu was a professor of physics and molecular and cellular biology at the University of California at Berkeley when President Barack Obama selected him to run the U.S. Department of Energy in 2009. Chu is a Chinese American, born in St. Louis, Missouri, in 1948. He won a Nobel Prize in Physics in 1997 for his research concerning the cooling and trapping of atoms with laser light. Chu believes that human-made global warming threatens civilization, and he says alternative energy researchers are fighting "a war to save our planet." In an interview with *National Geographic* magazine, Chu describes the government agency he runs:

The Department of Energy is an incredible science resource, and part of its mission is to provide the best scientific advice directly to policy makers—and that, fundamentally, is nonpartisan. . . . The Department of Energy is the biggest supporter of the physical sciences in the United States, but it also has a mission to take what is developed in national labs and universities and transfer this knowledge to applied research—research that will lead to really new ideas about sources of energy and ways of using our energy more efficiently.

Quoted in Michelle Nijhuis, "Our Energy Challenge," *National Geographic*, March 2009. http://ngm.nationalgeographic.com.

to a program called Coproduced, Geopressured, and Low Temperature Projects. This program is concerned with producing geothermal electricity from oil and natural gas wells. At these sites, an average of 10 barrels of hot water is pumped out of the ground with every barrel of oil. This wastewater, which must be removed from the oil, is considered a nuisance because drillers must safely dispose of it. However, the hot water can be put to productive use in binary cycle power plants. By working through oil-drilling facilities, the geothermal electricity will be coproduced with petroleum and natural gas.

At Southern Methodist University in Dallas, the Geothermal Energy Program was formed to conduct research into coproduction. Researchers at the university estimate that geothermal coproduction operations in the Texas Gulf region can produce up to 5,000 megawatts of power, equal to 10 coal-fired power plants. The State Energy Conservation Office of Texas explains the benefits associated with coproduction research: "This . . . could be a bonanza for the geothermal industry, while sparing the oil and gas industries the expense of disposal of these co-produced fluids. . . . Using tapped out oil and gas wells could greatly reduce the costs involved in exploration and drilling [for geothermal resources]."[10]

Research into coproduction is in its infancy. In 2010 there were only three research test projects in the United States. In Mississippi a company called Gulf Coast Green Energy is working with Southern Methodist University to generate 50 kilowatts of electricity from an oil well. In Louisiana researchers were trying to generate power from a natural gas well. And in Jay, Florida, a 1-megawatt project is being tested at the Jay Oilfield.

Hot Dry Rock

Not all geothermal plants rely on hot, underground water to provide steam. A technique called enhanced geothermal systems (EGS) pumps cold surface water at very high pressure into a drilled hole called an injection well. The cold water becomes superheated when it comes in contact with hot, dry rock that lies more than 5,000 feet (1,524m) below the surface.

In the EGS process, the cold water fractures solid rock and creates an underground reservoir from the cracks that extend in all directions from the drill hole, or wellbore. As the water flows through the cracks, it captures the natural heat in the rocks. The very hot water is then pumped out of a second wellbore, where it is converted into electricity by a steam turbine or binary power plant. The cooled water is then injected back into the first borehole, where it heats up again.

wellbore
A drill hole or bore hole created for the purpose of exploration or extraction of natural resources such as water, gas, or oil.

Geothermal in Australia

Enhanced geothermal systems projects are being developed in France, Japan, Germany, and in the United States in California, Arkansas, Nevada, and Oregon. But the world's largest EGS project is located in a central Australian region called Cooper Basin. Cooper Basin is ideal for EGS because 2 miles (3.2km) below the surface, the granite rock is heated to 455°F (235°C) by the earth's core heat. A Brisbane-based company, Geodynamics, has drilled a series of wells to tap into this heat with hopes of producing 500 megawatts of electricity by 2015. This is the amount of power from a typical coal-fired power plant, but the EGS produces electricity without any pollution.

Researchers believe that Cooper Basin alone could eventually produce 10,000 megawatts of electricity, enough to replace 20 coal-fired power plants. This will require a massive expansion of drilling efforts, the construction of several more power plants, and installation of transmission lines to take the power to Australian cities. According to Doone Wyborn, the chief scientist at the project, "Geothermal in Australia could potentially provide all the country's electricity needs for the next 100 years without any trouble."[11]

Predictions are not as optimistic in the United States. At current rates of research and construction, only 5 percent of the country's power is expected to be generated by geothermal resources by 2050. However, a report from the Massachusetts Institute of Technology estimates that if all available low-heat geothermal energy sites were tapped, they could produce more than 2,000 times the nation's electricity consumption. Chemical engineer Jefferson Tester of the Massachusetts Institute of Technology says, "This is a very large resource that perhaps has been undervalued in terms of the impact it might have on supplying energy to the U.S."[12]

Cutting Drilling Costs

Researchers believe that the best way to expand geothermal energy production is to lower the high costs of drilling deep boreholes into hard granite. A typical geothermal well 6 miles (9.7km) deep costs over $200 million. This makes EGS geothermal electricity more than 5 times as expensive as coal power. To lower EGS costs, Google.org, the philanthropic arm of the Internet search engine company, granted $4 million to Potter Drilling of California. The company is developing drilling techniques that melt rock instead of fracturing it—a technique that could substantially lower the cost of drilling. Most drillers use expensive diamond-tipped bits, which slow to a crawl when cutting through granite deep beneath the earth. If the drill breaks, the well must be abandoned, and all the money invested in drilling at that site is lost.

Potter Drilling founder Bob Potter, a scientist who helped develop the first atomic bomb in the 1940s, has invented a process called hydrothermal spallation. This technique does away with expensive drill bits and instead uses superheated water to fracture granite. Costs are further lowered because hydrothermal spallation is about five times faster than traditional drilling methods. With hydrothermal spallation, water is heated to 1,472°F (800°C) and forced at high pressure into the rock. The drilling method is explained on Potter Drilling's Web site: "The process starts by applying a high-intensity fluid stream to a rock surface to expand the crystalline grains within the rock. When the grains expand, micro-fractures occur in the rock and small particles called spalls are ejected."[13] The result is a borehole in the rock.

spallation

When used on rock, a process that causes particles to break free as extreme heat is applied over a small area.

The drilling of deep boreholes in hard granite requires expensive, diamond-tipped drill bits (similar to the one pictured here). At least one company is trying to lower the cost of drilling for geothermal energy by developing less costly drilling techniques.

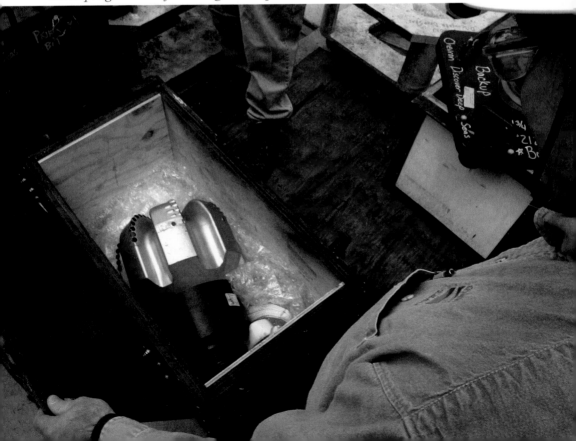

In a similar project, scientists at Argonne National Laboratory in Illinois are testing huge industrial lasers for drilling through softer rock such as shale, limestone, and sandstone. Like hydrothermal spallation, this process is fast and does not rely on expensive drill bits. In 2010 both the Argonne and Potter drilling projects were in the experimental stages. But researchers believe that by lowering the cost of geothermal drilling, EGS could someday be expanded at a far greater rate than would be possible with current technologies.

Solar Thermal Energy

While geothermal energy relies on heat from the earth, the sun's heat can also generate steam to drive turbines in power plants. This technology, called solar thermal energy, uses specialized mirrors, called heliostats, that track the sun's path across the sky every day. The

> **heliostat**
>
> An instrument in which a mirror is automatically and continuously moved so that it reflects sunlight in a constant direction.

heliostats form a large circle around a structure called a power tower receiver. Atop each power tower are mirrors that focus solar heat onto a boiler filled with water or a salt solution. When the sunlight hits the boiler, the liquid inside is heated and creates high-temperature steam. The steam is then piped to a turbine, which generates electricity.

Research on solar thermal energy began in 1978 with the construction of the National Solar Thermal Test Facility (NSTTF) in Albuquerque, New Mexico. The facility is run by Sandia National Laboratories, a research and development lab that is part of the DOE. By the mid-1980s the NSTTF had finished construction of a Central Receiver power plant, 222 heliostats, and a computer control system that generated electricity from the heat of the sun. In the decades that followed, scientists conducted research to improve heliostat and power tower designs as well as other technical issues concerned with making solar thermal energy cheaper and more efficient.

Research at the NSTTF resulted in usable commercial applications for solar thermal energy in 2007. Google.org joined with oil giants British Petroleum (BP) and Chevron to invest several hundred million dollars in a solar thermal energy project called the Ivanpah Solar Energy Generating Station. Construction on the Ivanpah station, which is located on a 4,000-acre (1,619ha) site in California's Mojave Desert, began in 2010. The generating station consists of 3 separate solar thermal power plants. The Ivanpah complex will generate enough electricity to power more than 140,000 homes. This will reduce carbon dioxide emissions by more than 400,000 tons (362,874 metric tons) per year, the equivalent of removing 70,000 cars from the road annually.

Environmental Concerns Lead to New Ideas

Solar thermal energy sites like Ivanpah are not without controversy, however. The Ivanpah site is home to rare plants and animals such as the desert tortoise, and environmentalists feared the facility would harm sensitive habitat. In addition, some say the 459-foot-tall (140m) power towers mar the visual beauty of the desert landscape. Environmental groups such as the Sierra Club, Defenders of Wildlife, and the Center for Biological Diversity generally favor solar energy projects but tried to block construction of the Ivanpah facility for these reasons.

Despite the objections, the California Energy Commission approved the project in March 2010. Terry O'Brien, the commission's deputy director, explained the need to balance renewable energy benefits with concerns about the desert ecosystem:

> [The Ivanpah Solar Energy Generating Station] will provide critical environmental benefits by helping the state reduce its greenhouse gas emissions, and these positive attributes must be weighed against the project's adverse impacts. It is because of these benefits and . . . the adverse impacts that global warming will have upon the state and our environment, including desert ecosystems, that staff believes it . . . [is appropriate] to approve the project.[14]

However, environmental concerns such as those expressed during the Ivanpah hearing process did not go unnoticed. Solar engineers have been searching for ways to utilize the power of the sun without marring valuable wildlife areas. One solution is to use rooftops of homes and commercial buildings. In cities such as Los Angeles and Phoenix, acres of rooftops could generate power right where it is needed. This is cheaper and easier than creating massive solar facilities in the desert and sending the power through hundreds of miles of transmission lines.

In July 2008 Southern California Edison became one of the nation's first electric utilities to harvest sunlight on rooftops of commercial buildings. The company did so using a technology called thin-film PV. Thin-film PV is made from a mineral called cadmium telluride, or cad-tel. This is about half the cost of the silicon used in traditional PV cells.

Traditional PV cells cost about $2 a watt, but experts say the cost of solar must fall below $1 a watt to be competitive with fossil fuels. At the end of 2009, First Solar, the world's largest thin-film manufacturer, achieved an 87-cents-per-watt manufacturing cost. This paved the way for solar expansion in the United States. Southern California Edison is planning to generate 250 megawatts of power (enough for 80,000 homes) by 2013 with First Solar thin-film PV panels placed on rooftops in San Bernardino County, east of Los Angeles. Pacific Gas and Electric, another utility in California, plans to produce electricity with thin-film PV placed on rooftops near the company's branch power stations, called substations, in northern and central California.

Catching the Sun

The search for less expensive, more efficient methods of producing solar-powered electricity has resulted in many other innovations. Solar thermal energy, for instance, is no longer limited to systems that rely on expensive central power towers. Researchers have combined technology from the 1800s with new technology to develop a solar electrical generating system. This system combines a heliostat with a Stirling heat engine, which was first conceived by British inventor Robert Stirling in 1816.

parabolic reflector

A concave reflective device used to collect or project light and other forms of energy waves.

The heliostat on a Stirling heat engine is a mirror shaped like a satellite dish, called a parabolic reflector. This follows the sun as it tracks across the sky and focuses the rays on a single spot. The concentrated sun rays are about 1,450°F (788°C), twice as hot as those in a standard solar cell. This heat drives a Stirling heat engine, which consists of a pair of pistons and cylinders. One of the cylinders in the engine is filled with hydrogen gas. When the hydrogen is heated, it expands, driving

The Stirling Energy Systems SunCatchers (pictured here at the Sandia National Laboratories Solar Thermal Test Facility in New Mexico) each have 40 precision mirrors attached to a parabolic dish. One SunCatcher can power about 20 homes.

the pistons up and down. The mechanical power is transferred to a generator, which produces electricity.

Stirling Energy Systems of Scottsdale, Arizona, built on this technology to develop a solar dish called a SunCatcher. SunCatchers have 40 precision mirrors attached to a parabolic dish. These mirrors focus the sun's rays onto a receiver. The receiver transmits the heat to a Stirling engine. Each SunCatcher can power about 20 homes. And there are plans to build 2 huge solar farms with SunCatchers in Southern

The Solar Eco-Fridge

Most people would not try to keep food cold by using the heat of the sun. But Emily Cummins of Keighley, West Yorkshire, England, has developed a small refrigerator that utilizes solar energy. In 2009 the 21-year-old student invented what she calls an eco-fridge. It is made from 2 cylinders, one inside the other. The outer cylinder has about 20 2-inch (5cm) holes and can be made from any common material, such as plastic or wood. The inner cylinder is metal, has no holes, and can hold perishable food such as meat, milk, or cheese. The gap between the inner and outer cylinders is filled with common insulating materials such as wool, sand, or dirt. This material is soaked with water, and the fridge is placed in the sun. The sun's rays cause the water to evaporate from the insulator material. This evaporation process pulls heat from the inner metal cylinder, which is cooled to about 43°F (6°C), preserving food placed inside. According to Cummins, "The simplest method of cooling something could be seen when you look at how we cool biologically—through sweating or evaporation. That idea led me to the design and the fridge was born."

After Cummins perfected the eco-fridge, she spent five months in Namibia, where she became known as the "Fridge Lady." Her invention changed the lives of thousands of poverty-stricken Africans who never had refrigerators, or even electricity.

Quoted in Chris Brooke, "Amazing Solar-Powered Fridge Invented by British Student in a Potting Shed Helps Poverty-Stricken Africans," *Daily Mail*, January 8, 2009. www.dailymail.co.uk.

California. These facilities will generate enough power for more than 1 million homes. When completed, the new facilities will more than double the amount of commercial solar electric power generated in the United States.

Stirling Energy founder Bruce Osborn has been working to perfect the design of the SunCatcher since he was an engineer at Ford Motor Company in the 1970s. His partner, engineer Chuck Andraka from Sandia National Laboratories, also has worked on solar Stirling systems for decades. Osborn and Andraka conducted research into other solar designs and noticed that most large-scale projects were custom-built on-site, a time-consuming, expensive process. To lower the price of Sun-Catcher solar farms, the engineers designed the unit so it could be made in a factory. As a result, a planned 900-megawatt Stirling Solar plant near San Diego will have 36,000 identical dishes, made cheaper by low-cost mass production. Maintenance is also less costly because a broken unit can be quickly replaced with a similar device.

SunCatchers also lower costs because they produce more electricity than other types of solar collectors. This was demonstrated on January 31, 2008, when the Stirling Energy team broke a 24-year-old solar efficiency record. Efficiency is measured by the ratio of the energy a solar dish produces to the total solar energy hitting the dish. The SunCatcher achieved a conversion rate of 31.25 percent, meaning almost one-third of the sun's energy was turned into electricity. This is about double the efficiency of standard solar cells.

Solar Flight

While millions of consumers will be receiving electricity from new solar test projects, Swiss pilot and inventor Bertrand Piccard is attempting what many believe is impossible. In 2009 Piccard unveiled an airplane he believes can fly for several days and nights without fuel. The 3,000-pound (1,361kg) aircraft, called the *Solar Impulse*, is powered by 11,000 photovoltaic cells mounted on its 211-foot (64m) wings, which are as long as those on a Boeing 747.

Solar flight is extremely dangerous because the energy gathered during the day must not only propel the plane, but also recharge the batteries to enable flying at night. Therefore, the pilot must have fully charged batteries when night falls in order to stay in the air until sunrise. To solve this problem, Piccard designed the PV cells on the *Solar Impulse* to charge batteries that power 4 electric motors. He hopes this will allow the sleek plane to fly to 27,000 feet (8,230m) and cruise at 45 miles per hour (72kph) day or night. Piccard worked with 50 engineers for 6 years to develop the $98 million plane. Commenting on the possibilities of solar flight, Piccard stated: "If an aircraft is able to fly day and night without fuel, propelled solely by solar energy, let no one come and claim that [it] is impossible to do the same thing for motor vehicles, heating and air conditioning systems and computers."[15]

The Promise of Solar and Geothermal Energy

The world is being changed every day by the promise of solar and geo-thermal energy. In China over 10 million people use solar hot water heaters. In India solar lanterns are lighting huts that have never seen a lightbulb. In Africa women are using solar refrigerators for the first time. And in the United States, vast solar projects are growing like wildflowers in the desert. As the twenty-first century progresses, light from the sun and heat from the earth are transforming society with the hope of a limit-less clean energy future.

Wind and Water

On November 8, 2009, wind power made headlines in Spain as high winds blew across the country. For more than 5 hours, 53 percent of the country's power was generated by the wind. This set a new record in wind energy production. The November winds were exceptional—Spain usually generates about 13 percent of its electricity from wind during the course of a year. But the record helped the Spanish government justify the $1.3 billion it had invested in wind power since 2007. Luis Atienza, president of Red Eléctrica, which runs Spain's electricity grid, said: "This makes us proud. There is no other country of our size which has completed and bettered a renewable energy production of over 50 per cent in such a timescale."[16]

While Spain's one-day record is impressive, wind energy proponents hope someday it will be commonplace for half of any nation's power to come from wind. For that to happen, however, new technologies need to be developed and perfected. In the United States, Energy Secretary Steven Chu does not think wind will ever provide half of the nation's power. In many places, such as the southeastern United States, land-based wind farms are not practical because the wind does not blow consistently and at high speeds. But Chu believes the United States could produce much more power from wind than the 1 percent it achieved in 2009. To this end, the DOE instituted a program called 20% Wind Energy by 2030. The program provides grants to researchers who devise ways to make wind turbines more efficient. The DOE is also looking to new frontiers, such as the oceans and skies, for wind farm placement.

Flying Electric Generators

The DOE reports that demand for electricity in the United States grows at about 1 percent a year, and this growth could be met with wind turbines. But, as critics of wind power point out, the wind does not always blow. Because it is inconsistent, reliable backup generators, such as natural gas and coal power plants, need to remain online. There is, however, a place where the wind blows consistently—high above the earth at altitudes of 1,600 to 40,000 feet (487m to 12,192m). In 2008 Cristina Archer, assistant professor of energy, meteorology, and environmental science at California State University at Chico, coauthored the first comprehensive study of high-altitude wind power. Archer determined that there is enough wind energy miles above the earth to meet global electricity demand 100 times over.

High-altitude winds can be harvested with machines called flying electric generators (FEGs). These devices have been tested in two forms; some are shaped like kites and others like helicopter-style rotorcraft. A San Diego–based company called Sky WindPower builds prototype rotorcraft machines consisting of four huge, 35-foot-long (10.7m) blades. A battery initially spins the blades, which then lift the device in the same way rotors raise a helicopter into the sky. Once in the air, wind takes over the work of the battery to spin the rotors, which

> **prototype**
>
> A full-scale functional model.

also act as turbines, turning an electrical generator located within the platform. The power is transmitted back to earth through a thick aluminum cable. The FEG can be kept on a steady course with a built-in global positioning system.

Company founders Bryan Roberts and David Shepard believe 43 "arrays," each made up of 600 FEGs, would generate enough electricity to power the entire United States. These would have to be located away from busy air traffic corridors, but Shepard says the arrays could be sent aloft in restricted air space where airplanes are not allowed to fly. The cost of such a system is still very high; Sky WindPower estimates the cost of a rotorcraft FEG to be about $2.26 million.

Kite-Style Generators

Smaller kite-style wind plants, called kitegens, are also designed to transmit energy through high-altitude winds. Parachute-shaped kitegens are lofted into the air like toy kites. The wind turns rotating cables, which spin a generator located on earth. Kitegens and FEGs are made to harvest winds from the jet streams, which are areas of very strong winds concentrated in narrow streams in the upper atmosphere. In the Northern Hemisphere, the jet stream typically blows from west to east at speeds up to 310 miles per hour (499kph). Flying electric generators, floated into such a powerful force, would have 80 percent efficiency, as opposed to

Cristina Archer Maps the Wind

In 2008 Cristina Archer coauthored the first comprehensive study of high-altitude wind power. She is a leading expert in ground-based, off-shore, and high-altitude wind mapping, and she used this knowledge to develop the first atlas of high-altitude winds. Archer's work is critical for those trying to find ways to loft flying electric generators into the jet stream to provide power for major urban areas.

Born in Italy, Archer was inspired to begin working in the environmental field by her middle school math and physics teacher, whom she describes as an eco-activist. Archer moved to the United States in 1996 to study meteorology because, at that time, the study of weather science in Italy was taught only in military aviation schools, and thus open only to men. In the United States Archer got involved in renewable energy because, as a meteorologist, she studied the winds. She says the move from meteorology to wind power and renewable energy was natural.

In 2010 Archer was assistant professor of environmental science, meteorology, and energy in the Department of Geological and Environmental Sciences at California State University at Chico. She was also a consulting assistant professor in the Department of Civil and Environmental Engineering at Stanford University. In addition to her academic work, Archer was an advisor to the Cleantech Innovation Center, established near Chico to spur research, training, and job creation in the green energy field.

the 20 to 30 percent achieved with ground-based wind machines. At this efficiency, high-altitude wind generators would be the cheapest source of power available—about half the cost of coal.

Another reason flying electrical generators lower costs is that they can be lofted over densely populated major cities, where power is most needed. The jet stream is particularly strong over New York City, making it a perfect place to float wind machines. Crowded cities in China and Japan have similar wind patterns. As Archer says, "The resource is really, really phenomenal. There is a lot of energy up there."[17]

Offshore Wind Farms

A great deal of energy is also available in the open oceans, where consistent winds blow at high speeds. In Germany researchers have taken advantage of swift ocean winds at the Alpha Ventus offshore wind farm, completed in late 2009. Located near the North Sea island of Borkum, the farm's 12 wind turbines are 34 miles (55km) from the German coast and planted in waters 100 feet (30m) deep. Although the Alpha Ventus wind farm supplies electricity for only 50,000 homes, German officials plan to build 40 offshore wind parks that will generate 10,000 megawatts of electricity by 2020. These farms, 7.5 to 124 miles (12km to 200km) offshore, will generate enough electricity for 8 million homes. An additional 10 wind farms are planned for placement in the Baltic Sea by 2030.

While the Alpha Ventus wind farm provides a working model for other such projects, it also highlights the problems associated with this technology. Offshore wind farms are very costly; the price tag of the Alpha Ventus wind farm came to $357 million. Part of the cost was for stone foundations for the turbines that had to be built into the seabed, a process that was highly technical and time-consuming. Repairs and routine maintenance are also costly because of the facility's location far from shore.

Researchers at the University of Maine are searching for ways to make ocean-based wind farms more practical by lowering construction and maintenance costs. One idea being pursued involves floating wind turbines. In October 2009 the DOE awarded an $8 million grant to the University of Maine's Deepwater Offshore Wind Consortium (DeepCwind). Researchers in the program are using the money to design and deploy floating turbines at an ocean wind farm called Stepping Stone. The goal is to have the turbines in place by around 2020. The floating turbines will be cheaper to put in place than offshore wind farms since they do not need stone foundations. Floating turbines only need to be moored to the seabed with anchors. And the floating wind generators can be towed to shore for maintenance and repair or redeployed in areas where the wind is stronger. Work will begin with the opening of the Offshore Wind Test Site, where two large, floating wind turbine prototypes will be designed, constructed, and deployed. This work will allow scientists to learn ways to design more durable turbines with lighter building materials that require less maintenance.

Floating wind turbines are advantageous because they can be used in very deep water, such as the Pacific Ocean off the West Coast of the United States where the seabed is more than 1,000 feet (304m) below the surface. According to the U.S. Department of the Interior, if floating wind farms were placed in the Pacific, they could generate 900 gigawatts of electricity every year, an amount equal to all currently installed U.S. electrical capacity.

Turbines at Germany's Alpha Ventus offshore wind farm harness the power of consistent, high-speed North Sea winds. Germany plans to build many more such wind farms in the coming decades.

Electricity from the Jet Streams

Flying electric generators are seen as promising because wind speed and reliability are far greater at high altitudes. This is particularly true in the subtropical jet stream along the U.S.-Mexican border and the polar front jet stream around the U.S.-Canadian border. It has been estimated that capturing just 1 percent of the available energy of these high-altitude winds would meet the electricity needs of the entire planet. A flying electric generator placed in the jet stream at an altitude of 15,000 feet (4,572m) above the earth could produce about 20 megawatts, 4 times the power of a ground-based turbine.

Floating Wind Generator Undergoes Testing

In 2010 Norway became the first nation to test a floating wind generator. Scandinavia's largest oil company, StatoilHydro, deployed a device, called Hywind, off Norway's southwestern coast. Hywind is designed for placement in waters 400 feet to nearly 2,300 feet (122m to 701m) deep. The $66 million wind machine holds a standard 152-ton (138-metric-ton), 2.3-megawatt turbine. But it is mounted 213 feet (65m) above the waves on a floating steel pole called a spar. The spar extends 328 feet (100m) below the surface and is anchored to the ocean floor by 3 cables that stabilize it and prevent the turbine from bobbing up and down excessively in the waves.

If this next-generation wind turbine can survive in Norway's frigid, turbulent North Sea, researchers say it will revolutionize offshore wind power. Hywind allows for exploitation of offshore wind power in countries with little available land and no shallow-water areas along their coastlines. This could make the machines perfect for populous countries such as Spain, Japan, and South Korea, where coastal waters are deep and ocean winds are abundant.

Floating Magnetic Turbines

Despite advances in offshore wind technology, most wind generators remain on land, where the wind is intermittent. This has led researchers to investigate ways to increase turbine efficiency so that machines can generate power even at low wind speeds. One way this can be done is by eliminating friction. In most wind generators the large rotor blades spin on ball bearings, which causes friction that slows the rotation of the blades and makes them less efficient. Bearings also wear out and need to be replaced, an expensive process.

Devices called maglev wind turbines do not need ball bearings, because they float on air a few inches above the ground. Maglev, short for "magnetic levitation," is based on a concept recognized by anyone who has ever tried to push two magnets together. A magnet has two ends, referred to as the north and south poles. When the north pole is placed near the south pole, the magnets attract. But if north is pushed against north, or south against south, the magnets repel.

Maglev wind turbines have vertically oriented blades arranged in a circle. The bottom of the blade structure is lined with magnets. The base has magnets with similar poles. The magnets repel, levitating the blade structure a few inches off the base. When the wind blows, the structure turns and produces electricity.

Without friction, maglev wind turbines spin with amazing ease. They are 20 percent more efficient than traditional wind machines and generate power at wind speeds as low as 4 miles per hour (6.4kph). The wind generators can also operate in winds blowing up to 50 miles per hour (80.4kph). And because they do not use ball bearings, maglev machines are 50 percent cheaper to maintain.

The world's largest maglev wind turbine is being constructed in central China at a cost of $53 million. However, a Canadian company called Enviro Energies produces small maglev turbines for home and business owners. The company's smallest machine, for a home rooftop, is about 5 feet high and 10 feet wide (1.5m by 3m) and costs around $12,000.

Underwater Wind Power

Researchers are pushing wind turbine efficiency to new heights. But as engineer David Watson states, "As cool as wind turbines are . . . I don't think wind energy will ever supply more than a small percentage of our (human race) electricity needs. You just can't get that much energy out of [wind machines]. You'd have to have them just about everywhere."[18] However, wind is not the only element that can spin blades to drive turbines. A company called Hydro Green Energy is building turbines for the water. These machines are called hydrokinetic because they generate power from the kinetic energy of moving water. When river water flows downstream, it spins the propellers of the turbine.

Hydrokinetic turbines are considered superior to large hydroelectric dams for several reasons. They can be built in factories and placed in rivers at a fraction of the cost of a dam. In addition, river turbines do not have the environmental impact of dams, which destroy river ecosystems.

hydrokinetic

Relates to fluids in motion or the forces that produce or affect such motion.

In early 2009 Hydro Green Energy deployed the nation's first commercial hydrokinetic turbine in the Mississippi River near Hastings, Minnesota. The 35-kilowatt turbine is placed near a hydroelectric dam. This location allows the turbine to feed the power it generates into the existing electrical grid. The hydrokinetic turbine is expected to increase the capacity of the existing power plant by 2.5 percent. While that number is small, Mark Stover, a vice president at Hydro Green Energy, describes the advantages of hydrokinetic turbines: "We don't require massive dam construction, we're just using the natural flow of the stream. It's underwater windpower if you will, but we have 840 or 850 times the energy density of wind."[19]

The main problem experienced by designers of hydrokinetic turbines has nothing to do with generating energy. There are many regulations that govern what type of devices can be placed in rivers. And low-impact turbines are classified in the same category as large, environmentally destructive hydroelectric dams. The Federal Energy Regulatory Commission oversees all projects where power is produced with water, and turbine projects can take years to be approved by the agency. However, in 2008 Philip Moeller, commissioner of the Federal Energy Regulatory Commission, took steps to streamline the approval process for the Hydro Green Energy project and others like it. Moeller stated, "I am thrilled to support today's historic order that allows for harnessing more power from the Mississippi River. . . . I hope this is the first of thousands of similar projects that produce clean and renewable power from in-stream flows at existing dams."[20]

Deep Sea Vibrations

While hydrokinetic energy can be harnessed to create power, it also destroys machinery. And the destructive power of moving water has long been a problem for offshore oil-drilling platforms like those in the Gulf of Mexico and elsewhere. Drilling platforms are held in place by long, cylindrical cables. These anchor the floating platforms to the seabed. However, when ocean current flows over a cable, a small vortex, or swirling eddy, is formed. These vortices cause vibrations that destroy the oil platform's moorings over time.

Michael Bernitsas, an ocean engineer at the University of Michigan, spent many years working for the oil industry and searching for ways to suppress the vortices. However, Bernitsas came to understand that the vortices could be used for something good: "It dawned on me that we should enhance the vibrations and try to harness the energy."[21]

One of the companies involved in water-to-energy research demonstrates its tidal water hydrokinetic system, which will convert moving water to usable electric energy. Hydrokinetic turbines are considered a good alternative to large hydroelectric dams.

Bernitsas developed a prototype technology called the Vortex Induced Vibrations Aquatic Clean Energy (VIVACE) system. In its simplest form the VIVACE consists of small, round cylinders a few inches across, suspended on a spring. The cylinders have small fins, like fish, that help them capture more energy from the vortices. As the currents swirl around the finned cylinders, they move up and down. The mechanical energy is transferred to generators to make electricity. Bernitsas believes the concept could be used to create arrays that could generate one gigawatt of energy, about one-third the energy created by a large nuclear power plant.

vortex

A spinning, often turbulent, flow of fluid; the kinetic energy of vortices may be utilized to generate power.

The VIVACE system is seen as advantageous because it works in low-speed currents. Hydrokinetic generators like those used in the Mississippi River need a current flowing at least 7 miles per hour (11kph), whereas most ocean currents are about 3.5 miles per hour (5.6kph). The VIVACE system works in currents flowing a little over 1 mile per hour (1.6kph). As Bernitsas says: "There is a huge amount of hydrokinetic energy in currents but a lot of that we cannot harness with the present technology and that's where my device comes in, to extract energy at speeds down to 1 [mile per hour]. It taps into a new energy source."[22] The National Science Foundation, the U.S. Navy, and the DOE have together contributed $2 million to allow Bernitsas to continue work on the concept.

Ocean Thermal Energy Conversion

Ocean engineers like Bernitsas traditionally study the effects of the ocean environment on ships and marine structures like oil platforms. But the latest cutting-edge science in ocean engineering involves harvesting the nearly infinite thermal energy produced by sun and sea. The leading engineer in this field, John Piña Craven, is mainly interested in the large difference between ocean water temperatures found 2,000 feet (610m) below the surface and the much warmer water above it. The temperature difference that exists between deep and shallow waters can be used to run a heat engine in a process called ocean thermal energy conversion (OTEC).

Craven founded the world's leading institution for OTEC in 1974, the state-owned Natural Energy Laboratory of Hawaii on Keahole Point, near Kona. In that area the deep ocean water is around 43°F (6°C). However, the surface waters around Kona are reliably heated by the sun. This solar energy brings surface water temperatures to 76°F to 82°F (24.5°C to 27.5°C) year round.

Craven says that the conversion process could provide more than 300 times humanity's electrical needs. To utilize that energy, the warm surface water is pumped into a heat exchanger to boil propane, which has a very low boiling point. The boiling propane creates steam, which passes through a turbine and causes it to rotate, generating electricity. After the propane passes through the turbine, it moves into a tank cooled by the cold water pumped from deep in the ocean. This cools the propane back to liquid form so it can be cycled back through the system.

The Natural Energy Laboratory of Hawaii successfully produced electricity from OTEC in 1979. But as oil prices declined during the early 1980s, so did the interest in OTEC. Electricity produced with the experimental technology was more expensive than oil power. However, research continued and in 2008 aerospace giant Lockheed Martin broke ground on an OTEC pilot project in Hawaii. The company estimated the 10-megawatt OTEC power plant will cost $400 million and generate 30 times more electricity than a coal-fired power plant.

In addition to electricity, OTEC can produce other valuable commodities that are in great demand. The cold water running though the pipes can be used to produce cost-free air-conditioning. At the Natural Energy Laboratory of Hawaii, the cold seawater used for air-conditioning saves the facility nearly $4,000 per month in electricity cost. In addition, the pipes "sweat" with condensation, which yields a large supply of freshwater for drinking and irrigation. Commenting on the promise of OTEC, Joseph Huang, a senior scientist for the National Oceanic and Atmospheric Administration, states: "The potential of OTEC is great. The oceans are the biggest solar collector on Earth, and there's enough energy in them to supply a thousand times the world's needs. If you want to depend on nature, the oceans are the only energy source big enough to tap."[23]

The Potential of Water and Wind Energy

The technology exists to harvest energy from the oceans, but it needs to be improved before it can replace cheap and abundant fossil fuels. Like other forms of water power, OTEC remains a small percentage of the total renewable energy picture. However, growing numbers of scientists like Craven, Bernitsas, and Archer are conducting research every day to find new ways for water and wind to provide clean energy for generations to come.

A New Generation of Biofuels

The United States produces more biofuels than any other country and almost all of it is in the form of ethanol made from corn. In 2009, the United States produced 9 billion gallons (34 billion L) of ethanol. This amounts to about 2 percent of the nation's transportation fuel but about 30 percent of the entire U.S. corn crop. If all U.S.-grown corn were used for ethanol instead of food, it would only supply about 7 percent of the nation's needs. For these and other reasons, researchers are developing alternatives to corn-based ethanol.

Venture capitalist Vinod Khosla is betting billions on new biofuel technologies. Khosla is a founder of Sun Microsystems, was an early investor in Google, and in 2010 was the 317th richest person in America. He founded Khosla Ventures in 2004 to invest in clean energy technology, or cleantech. Since that time he has directed hundreds of millions of dollars to a dozen different biofuel companies. While the companies are diverse, they have one thing in common—their scientists are searching for new and revolutionary ways to produce transportation fuel from plants. Khosla's main interest is creating new biofuels that can replace fossil fuels within 5 to 7 years of their initial deployment. As Khosla says, "I believe in technologies that can compete in the marketplace. Otherwise, it's just toys you're dealing with. Not solutions."[24]

First- and Second-Generation Biofuels

Biofuels are created from the earth's biosphere. The biosphere is the name of the thin surface layer of earth where the biomass, or all living matter, exists. While the biosphere represents only a tiny sliver of the earth's mass, it contains vast amounts of energy stored in plants and other living matter. And the store of biomass is constantly replenished by the sun. Scientists are looking for new ways to harness the energy that is stored in the biomass and convert it to liquid or gaseous fuels.

The most common method of converting plant energy into fuel is the process of turning corn, soy, palm oil, and sugarcane into

Cornfields surround an ethanol production plant in South Dakota. Nearly all of the ethanol produced in the United States comes from corn. Researchers are developing biofuels that do not depend on corn and other important food crops.

ethanol. The use of food crops for fuel is referred to by scientists as first-generation-biofuel production. First-generation biofuels are considered problematic. In a world where 15 million children die of hunger every year, some feel it is immoral to use food crops to fuel cars and trucks.

Second-generation biofuels are considered superior because they are produced from waste products and nonfood crops. One of the companies Khosla is backing is a second-generation producer. Range Fuels in Broomfield, Colorado, founded by inventor Bud Klepper in 2000, has successfully converted wood waste, switchgrass, municipal solid waste, hog manure, and even olive pits into biofuels. These raw materials are called feedstock. They are cellulosic—that is, they are composed of cellulose, the tough structural material found in the leaves, stalks, and roots of green plants as well as paper pulp, hog manure, and other organic substances.

Cellulose cannot be digested by humans, but it can be converted to fuel. And many nonfood plants produce much higher yields of ethanol than corn. However, corn is easier and less expensive to convert to alcohol because of the way ethanol is made. In the production of corn ethanol, enzymes are added to the plant fiber, which breaks down the starch in the corn kernel, yielding simple sugars. Yeast, a microorganism classified as a fungus, is added to the mix. The yeast eats the sugar and excretes alcohol (ethanol) and carbon dioxide. The yeasts that convert simple sugars to ethanol are well understood. They have been used to make beer, wine, and liquor for centuries. And the process is cheap. According to the DOE, it costs about $1.10 to produce 1 gallon (3.79L) of ethanol from corn.

> **cellulose**
>
> The tough structural material found in the leaves, stalks, and roots of green plants as well as paper pulp, hog manure, and other organic substances.

A More Efficient Process

Because cellulose is woody and fibrous, it is much tougher than corn to convert into sugar. In order to do so, enzymes are added to cellulose that break down plant cell wall tissues. These enzymes cost 30 to 50 cents per gallon, compared to 3 cents per gallon for corn. To bring costs down, the traditional fermentation process used with corn must be discarded when producing cellulosic ethanol.

> **fermentation**
>
> The conversion of carbohydrates or other substances into alcohol.

Klepper patented what is called the thermochemical conversion process, which uses more efficient methods to produce cellulosic biofuels. Instead of breaking the cellulose into sugar molecules, the raw material is converted through various combinations of temperature, pressure, water, oxygen, and chemicals. This process transforms the cellulose into synthesis gas, or syngas. The syngas consists of carbon monoxide, carbon dioxide, and hydrogen. This is fed into a special fermenter, where the gas is mixed with bacteria. These microorganisms eat carbon monoxide, carbon dioxide, and hydrogen and excrete ethanol and water. The ethanol is separated from the water and used for fuel.

The Range Fuels process generates no toxic waste and very few greenhouse gases. The company is planning to produce 100 million gallons

Flying High on Algae

In February 2008 a Virgin Atlantic Boeing 747 took off from Heathrow Airport in London. The huge jetliner landed in Amsterdam about an hour later. While dozens of jets fly between these two cities every day, the Virgin Atlantic airliner was the first to do so with one of its four fuel tanks carrying a 20 percent mix of plant-based fuel. The biofuels were made from oil processed from coconuts and the Brazilian babassu plant. After the flight, British billionaire Richard Branson, who owns Virgin Atlantic, spoke to the press: "What we're proving today is that biofuel can be used for a plane. Two years ago, people said it was absolutely impossible." Critics point out that only 5 percent of the total fuel load consisted of biofuels and that it would take 3 million coconuts had the flight been made entirely on biofuel. Branson understands this, and since the flight, Virgin Atlantic has been working with the Boeing Company, the world's leading manufacturer of commercial airliners, to develop algae as a source of aviation biofuel. Branson calls algae "the jet fuel of the future."

Quoted in Laura Blue, "Can Airplanes Fly on Biofuel?" *Time,* February 25, 2008. www.time.com.

(378 million L) of fuel a year while consuming one-quarter of the water used by corn ethanol plants with similar yields. Klepper estimates Range Fuels could someday produce 2 billion gallons (7.57 billion L) of cellulosic ethanol.

Range Fuels plans to use wood waste from logging operations as its primary feedstock. But the plant can also use a 10-foot-tall (3m) grass called switchgrass, which grows wild on the marginal land around food crops. According to biology researcher Sandy McLaughlin, who leads the U.S. government's switchgrass research effort at Oak Ridge National Laboratory:

> Producing ethanol from corn requires almost as much energy to produce as it yields, while ethanol from switchgrass can produce about 5 times more energy than you put in. When you factor in the energy required to make tractors, transport farm equipment, plant and harvest, and so on, the net energy output of switchgrass is about 20 times better than corn's.[25]

Third-Generation Fuels

Second-generation-biofuel production relies on an improved manufacturing process that can turn woody plants into fuels. To create a third generation of biofuels, researchers are seeking to improve the feedstock. This is done by genetically modifying common plants to produce more oil or other substances that can be turned into fuel.

Genetic modification, or genetic engineering, uses technology to alter the genes of microorganisms, plants, and animals. In this process the genetic material from one organism is transferred to another. This allows the desirable traits of two organisms to be combined. Since the 1990s many types of food have been genetically modified. This has been done to increase resistance to pests and disease, improve flavor, delay ripening, or increase shelf life. For example, certain strains of russet potatoes have been modified with a gene to produce a natural toxin that protects them against the potato beetle.

Some in the biofuels industry view genetic modification as a positive step toward producing better feedstock. For example, researchers at the Biotechnology Foundation Laboratories of Thomas Jefferson University in Philadelphia have discovered that they can genetically alter the leaves of the tobacco plant to produce great amounts of oil. The oil can be extracted and used in biodiesel, and the leaves left as waste can be fermented to make ethanol.

In most plants, biofuel oil is extracted from the seeds. While tobacco plants produce very oily seeds, they do not make enough to use for mass production of biofuels. However, researchers have been able to genetically modify tobacco to shift the oil production into the plant's large, abundant leaves. Scientists Vyacheslav Andrianov and Nikolai Borisjuk, who achieved this goal, write:

> In the search for alternative biofuel plant resources, tobacco has been largely overlooked as it is considered primarily as an expensive crop grown for smoking. When grown for energy production instead of smoking, tobacco can generate a large amount of inexpensive biomass more efficiently than almost any other agricultural crop. By generating both biofuel oil and ethanol, tobacco has the potential to produce more energy per hectare than any other non-food crop.[26]

This discovery could help tobacco farmers who have been struggling over the years as fewer people take up smoking. Researchers say that acre for acre, genetically modified tobacco can produce twice as much oil as soybeans.

Oil from Algae

Whatever the benefits of genetically modified feedstock, plants like tobacco are grown with pesticides, fertilizers, and agricultural machinery that guzzles fossil fuels. That is why some see algae farms as the ideal way to produce carbon-neutral, third-generation biofuels from microscopic organisms.

Some species of algae contain over 50 percent oil, and the microscopic organisms can be genetically modified to produce even greater quantities. Growing algae does not require expensive farmland, because the organisms can be grown at sewage treatment plants in wastewater.

Tobacco plants (pictured) offer another promising source for biofuel production. Researchers say that this can be accomplished with the help of genetic engineering.

And while 1 acre (0.4ha) of soybeans can produce 50 gallons (189L) of oil, a pool of algae that size could produce 6,000 gallons (22,712L). However, growing algae and harvesting it by removing it from water is difficult and expensive.

Algae farms use ponds and column-like water tanks called bioreactors. The floating algae is suspended in water. About half the cost of growing algae comes from providing a steady supply of food and water. To harvest algae, electric-powered rotating devices called centrifuges spin rapidly to separate the plant from the water. These processes produce algae biodiesel that costs about $56 a gallon ($14.77 per liter). This was about 18 times the cost of petroleum diesel fuel in 2010.

Two engineers at Kansas State University, Wayne Yuan and Z.J. Pei, are searching for ways to produce algae biodiesel for around $5 a gallon ($1.31 per L). This will require genetically modifying an oil-rich algae species that does not float but will instead grow on a solid surface. Yuan and Pei envision using large platforms placed in the ocean, where the algae would receive a steady supply of water and nutrients.

To advance their concepts Yuan and Pei are studying mechanisms algae use to attach to various surfaces. The scientists also need to know the type of materials algae prefer and the surface textures that encourage the algae to bloom and grow. In 2010 Yuan and Pei were using a $98,560 Small Grant for Exploratory Research from the National Science Foundation to find answers to these questions.

Fourth-Generation Biofuels

In 2009 the DOE announced $786 million in grants for the creation of what Energy Secretary Chu calls "third-generation biofuels like green gasoline, diesel, and jet fuels."[27] The DOE is also interested in fourth-generation biofuels, which are called carbon negative. Rather than produce CO_2, the fourth-generation fuels remove carbon dioxide from the atmosphere when they are being created. Carbon-negative fuels would allow humans to reverse global warming while producing fuel for their cars and trucks. To achieve this goal, researchers are focusing on fuel-making microbes that can be synthesized in the lab.

J. Craig Venter is the leading proponent of fourth-generation biofuels. He is an influential biologist best known for his work mapping the human genome in 2000. The genome is the entirety of an organism's hereditary information encoded in the DNA of its cells. DNA is often compared to a blueprint because it contains the instructions to make the proteins that build and run an organism. Understanding the genome allows scientists to perform genetic modifications.

Venter founded the company Synthetic Genomics in 2005 with Nobel Prize laureate Hamilton Smith. The company is using a process called synthetic genomics to create the world's first human-made species. It is a synthetic microbe called *Mycroplasma laboratorium* that could lead to an endless supply of biofuel. To create the synthetic organism, Venter's team combines slices of DNA into a single organism. The synthetic organism is engineered to consume CO_2 and secrete gaseous or liquid biofuels as waste products. Venter claims *Mycroplasma laboratorium* can make either hydrogen or ethanol, stating: "Because we actually have to feed [the microorganism] concentrated CO_2, we can take CO_2 . . . from power plants, cement plants and other places. . . . We want to take all that waste product and convert it into fuel."[28]

Venter says if his project is successful, people will be able to grow *Mycroplasma laboratorium* in their backyards to produce their own fuel. He sees millions of "algae micro refineries," producing liquid fuel that "can basically be used right away as biodiesel."[29] Producers would only need hardware such as sinks, filters, barrels, and pipes to produce backyard biofuel. If Venter can find a way to use a microscopic organism to create hundreds of millions of gallons of fuel from CO_2, it could significantly reduce the greenhouse gases in the atmosphere.

Bug Oil 2.0

Khosla is also looking at microscopic organisms in the quest for genetically modified oil producers. In 2008 Khosla Ventures invested $20 million in a company called LS9 that is hoping to create genetically modified insects. These creatures will excrete oil after feeding on agricultural waste such as woodchips or wheat straw. LS9's oil producers are much smaller than average insects, one-billionth the size of an

DNA

Deoxyribonucleic acid, contains the biological instructions that make each species unique. DNA, along with the instructions it contains, is passed from adult organisms to their offspring during reproduction.

J. Craig Venter

J. Craig Venter is one of leading scientists in the field of genomic research—and he is now applying his understanding of genetics to create synthetic microbes that can produce an endless supply of biofuel. Born in Salt Lake City, Utah, in 1946, Venter was not a very good student in middle school. He moved to San Mateo, California, during high school and upon graduation was drafted into the military and sent to fight in Vietnam. There he worked in an intensive care ward in a field hospital. Tending wounded and dying soldiers, Venter was inspired to study medicine and scientific research at the University of California at San Diego.

In 1984 Venter joined the National Institutes of Health, where he developed a revolutionary new strategy for discovering genes called expressed sequence tags. In 1992 he founded the Institute for Genomic Research, where he led a team that decoded the genome of the bacterium *Haemophilus influenzae*, which causes pneumonia and other diseases. During this work, Venter invented a new method of genome identification called the genome shotgun technique. In 1998 Venter founded Celera Genomics to sequence the human genome using this new technique. In February 2001 the completed sequence of the human genome was published in the journal *Science*. In 2006 Venter founded the J. Craig Venter Institute to explore new biological sources of energy. Venter believes the consequences of global warming will be catastrophic unless new energy sources can be found to replace fossil fuels as soon as possible. By using the pioneering research of genomic science, Venter is working to invent oil-creating microbes to forestall a global crisis.

ant. They start out as yeast, but LS9 microbiologists modify them by redesigning their DNA. The bugs excrete a type of fatty acid that is only slightly different than crude oil on a molecular level. Company founder Greg Pal calls this substance the organisms excrete "Oil 2.0." It is carbon negative and interchangeable with oil as soon as it is excreted, requiring no further processing. As environmental journalist

Chris Ayers explains: "Using genetically modified bugs for fermentation is essentially the same as using natural bacteria to produce ethanol, although the energy-intensive final process of [ethanol] distillation is virtually eliminated because the bugs excrete a substance that is almost pump-ready."[30]

While Oil 2.0 holds promise, LS9 can only produce about 265 gallons (1,000L) at a time in its fermenting machine. To produce enough to fill the gas tanks of U.S. drivers for a week would require a factory that covered an area the size of Chicago. This has not deterred Pal from his work, as he explains: "I have two children, and climate change is something that they are going to face. The energy crisis is something that they are going to face. We have a collective responsibility to do this."[31]

Biomass Cogeneration

Most cutting-edge biomass science concerns biofuels. But trees, wood waste, paper, and cardboard can also be used to generate electricity. This biomass can be burned directly in an industrial furnace to create steam that turns a generator and produces electricity.

The global pulp and paper industry has substantially increased its use of woody biomass for energy in recent years. Numerous pulp and paper plants have installed new equipment to switch from fossil fuels to woody biomass. As a result, global consumption of biomass in the wood and paper products industry increased 51 percent between 2006 and 2009. The leading biomass-consuming countries, not surprisingly, are those with large areas of forests, including Canada, Brazil, and Sweden.

In the United States, Nippon Paper Industries USA broke ground on a biomass plant in 2010 at its Port Angeles, Washington, factory. The $50 million facility, called a cogeneration plant, is designed to make electricity from paper factory waste and wood waste, called slash, left behind by logging operations. The company plans to use the power to run the paper mill while selling the excess energy to the local power company.

In South Carolina, Peregrine Energy Corporation constructed a $135 million, 50-megawatt cogeneration plant to provide power to 14,000 homes. The plant replaced coal-fired boilers with renewable energy from the region's logging operations. Traditionally, when forests are thinned or harvested, woody debris from tree trunks, limbs, and tops is left on the

Fuel pellets made from wood waste (pictured) have replaced coal as a fuel source at a Wisconsin paper mill. Paper and pulp mills worldwide have substantially increased their use of woody biomass for energy.

ground to rot. This dying vegetation attracts insects and unhealthy microorganisms that cause disease in the healthy trees nearby. The rotting wood also releases methane, a greenhouse gas, into the environment. Peregrine's biomass plant produces low amounts of CO_2 when burning this waste. However, researchers have determined that the facility is fourth generation, or carbon neutral, because it prevents the release of methane from decaying wood.

Cheaper than Solar

The biomass industry is one of the fastest-growing sectors of the renewable energy market. Ethanol production alone more than tripled worldwide between 2000 and 2009. This was based largely on first-generation technology. However, as second-generation biomass methods are perfected, researchers predict that ethanol production will double in the United States alone by 2020. Similar growth in ethanol production is expected in China and India. And if third- and fourth-generation biofuel production can be expanded, 100 percent of the world's cars, trucks, and airplanes might someday run on biofuels.

Biomass production is one more way that humanity is utilizing the power of the sun. And as Amol Deshpande, a partner with the venture capital firm KPCB, points out, it is cheaper to harvest the sun's energy with plants than with solar panels. Deshpande says: "I tell my partners . . . biomass is fourth generation solar energy. Capturing solar energy in the form of biomass and using that resource in a distributed way to make energy should be cheaper than solar for many years to come."[32] While challenges lay ahead, the most promising advances in renewable energy technology may be found in the new generation of biofuels being readied for the coming decades.

Hydrogen and Fuel Cells

Australian electrochemist John O'Mara Bockris coined the term "hydrogen economy" in 1963. Bockris envisioned a day when all transportation and electricity needs would be taken care of by hydrogen-powered fuel cells. Few people gave much thought to the concept of a hydrogen economy in the decades after Bockris wrote about it. Fossil fuels were cheap, and global warming had not yet been identified as a problem. However, in recent years the idea of a hydrogen economy is being viewed in a new light because of the advantages fuel cells have over traditional energy sources.

Fuel cell batteries are nonpolluting—the only by-products of hydrogen fuel cells are oxygen and water. If hydrogen is produced with renewable electricity, no greenhouse gases result. And hydrogen can be produced anywhere there is electricity and water.

Technological Hurdles

Despite the benefits of fuel cells, several major technological hurdles have prevented the hydrogen economy from taking hold. The first concerns hydrogen itself. Unlike oil, natural gas, and coal, which are extracted from the ground, pure hydrogen does not naturally exist in any useful quantities. About 90 percent of the hydrogen in the United States is derived from natural gas in a complex process called steam-reforming. Steam reforming requires more energy from the natural gas than is available in the hydrogen that results. Hydrogen can also be created from water using a technique called electrolysis. This process requires three times more energy than it produces.

There is currently not enough electricity generated in the entire United States to produce hydrogen on a massive scale. If all American vehicles were converted to fuel cells, the equivalent of a thousand nuclear power stations would have to be constructed to supply electricity for hydrogen production. This is more than double the number of working nuclear plants in 2010.

Other issues with hydrogen concern storage and transportation. Hydrogen is the lightest gas in the universe, and it floats into space unless it is tightly contained. Unlike oil, which is easily transported in metal barrels, hydrogen must be stored in expensive, leak-proof storage tanks. Hydrogen is also bulky, as chemistry professor Lev Gelb explains: "If you had a kilogram of hydrogen . . . you'd have to store it in about 100 big balloons, if you can picture that. A kilogram of gasoline, on the other hand—that would [fit into] a small container." This means hydrogen must be compressed to fit in tanks, a process that requires natural gas, prompting Gelb to state: "The case has been made persuasively that you'd be better off just burning the natural gas, rather than going to the trouble of producing hydrogen from natural gas and going through all the problems associated with its storage and transport."[33]

The final problem with fuel cells is the cost of a necessary element called a catalyst. A catalyst is something that accelerates a chemical reaction. In a fuel cell platinum is used as a catalyst to split hydrogen from oxygen molecules. Platinum is a nonrenewable precious metal; in 2010 an ounce of platinum cost around $1,660. That pushed the price of a 200-horsepower fuel cell like those used in experimental fuel cell vehicles to about $70,000.

> **catalyst**
>
> A substance that accelerates a chemical reaction.

Fuel Cell Functions

When Bockris conceived of the hydrogen economy, fuel cells were new inventions. Developed by the National Aeronautics and Space Administration (NASA) for space exploration in the early 1960s, the hydrogen-powered fuel cell played an important role when the United States sent astronauts to the moon in 1969. The cells provided electricity and potable water for the entire eight-day mission. Today fuel cells remain an integral part of space travel. The Space Shuttle orbiter has three fuel cell power plants on board that transform hydrogen and oxygen into electrical power, water, and heat. And NASA is exploring next-generation fuel cell concepts for future space vehicle applications. One example is the Mars Flyer, a fuel cell–powered aircraft designed to collect images and scientific data while flying over the Valles Marineris canyon at the Martian equator.

The NASA fuel cell is called PEM, which stands for polymer electrolyte membrane or proton exchange membrane. It is similar to fuel cells used on Earth for transportation and electricity generation.

⚛ The Home Energy Station

While Honda was developing the Clarity FCX fuel cell vehicle, the company's engineers were also working on ways to deliver hydrogen to owners of the car. Hydrogen fueling stations, of which there are few, can cost $2 million each. Honda decided to try something different. The company built an experimental home equipped with a fueling station in Torrance, California, in 2005. The home's fueling station produces hydrogen from natural gas. The hydrogen from this process can be used to power a Clarity along with a small fuel cell electrical system for the home.

Natural gas already heats more than half of the homes in the United States. This makes it a convenient feedstock for hydrogen. Although producing hydrogen from natural gas is costly, the fueling station also heats the home and provides hot water and electricity, which offsets the cost. Making hydrogen from natural gas still releases carbon dioxide. But Honda says that CO_2 output from the system is 30 percent lower than for a household that uses grid electricity and drives a gasoline-powered car.

The science behind PEM fuel cells is complex and takes place at a molecular level. Hydrogen gas is fed into an anode, or a positive terminal that flows into the fuel cell. Oxygen is fed into a cathode, or negative terminal, on the other side of the cell. At the anode, a chemical reaction is accelerated by a chemical, called a catalyst, made from platinum. The catalyst causes the hydrogen to split into positive ions and negatively charged electrons. The PEM allows the positively charged hydrogen ions to pass through to the cathode. The negatively charged electrons are forced to travel to the cathode along an external circuit, creating electrical current. After generating electricity, the hydrogen molecules combine with oxygen to form water, which flows out of the cell. While fuel cells are highly technical, they work like regular batteries, except they never wear out and do not need to be recharged. As long as the cell receives fuel, it will generate electricity.

ion

An electrically charged atom.

Fuel Cell Vehicles

The biggest promoters of PEM technology are major automobile companies, which have invested billions of dollars in fuel cell vehicles (FCVs) since 2000. Promoters of fuel cell cars, trucks, buses, and bikes believe PEMs have many advantages over traditional internal combustion engines fueled by gas and diesel. Fuel cell vehicles using hydrogen produced from natural gas reduce greenhouse gas emissions by about 50 percent. When hydrogen comes from clean energy, like solar electrolysis or biomass, greenhouse gas emissions are zero. And fuel cells are about 68 percent efficient, in that they turn more than two-thirds of the hydrogen energy into power. This compares favorably with internal combustion engines, which only utilize about 15 percent of the energy in gasoline. (The rest of the energy is lost to engine and driveline inefficiencies and idling.)

In 2010 several models of experimental FCVs were being produced by major automobile manufacturers. They included the Daimler F-Cell, the Hyundai Tucson FCEV, and the Volkswagen Tiguan HyMotion. However, the only vehicle available to the public, in limited numbers, was the Honda FCX Clarity. Like all FCVs, the Clarity is whisper quiet, emitting a soft electrical hum rather than the loud noise of an internal combustion engine. But only about a dozen people were able to lease the sleek, futuristic Clarity, most of them government officials and celebrities. These people paid about $600 a month to drive the Clarity. But because the vehicle was a hand-built prototype, the actual price of each car was about $1.5 million.

Most analysts believe that in order for FCVs to replace traditional cars, the price of individual vehicles would have to be reduced to at least $35,000. While Toyota has not been able to break that price threshold, the company announced it will begin selling a $50,000 FCV in 2015.

In order to cut costs, Toyota reduced the amount of platinum used in the fuel cell. Current fuel cells use about 1 ounce (28g) of platinum, but Toyota plans to cut that by two-thirds. The company also believes that mass-producing the car on a large scale will make it cheaper. Commenting on Toyota's plans for a cheaper FCV, green energy journalist Laura Sky Brown explains, "In the same way as the personal computer, it wouldn't be surprising to see this technology start out prohibitively expensive and rare, and work its way to becoming ubiquitous and affordable."[34]

Honda Motor Company workers in Japan place a hydrogen tank on a new FCX Clarity. Honda has begun commercial production of the zero-emission hydrogen fuel cell car but because of cost early versions had limited appeal.

Biological Hydrogen

Before FCVs can replace other types of vehicles, cheaper and greener methods for large-scale hydrogen production need to be devised. To lower costs and increase availability, researchers are working to create hydrogen through biological means, producing what is called biohydrogen. This process employs photosynthesis and microorganisms like bacteria. During photosynthesis, plants use the energy of sunlight to convert carbon dioxide in the atmosphere into sugars. The sugars provide usable fuel for a plant to grow and reproduce. However, a unique type of green algae called *Chlamydomonas reinhardtii* contains an enzyme called hydrogenase. This enzyme produces trace amounts of hydrogen during photosynthesis. Like all other algae, the organism has no roots and can be grown in water almost anywhere. But when *Chlamydomonas reinhardtii* is exposed to excessive sunlight, some of the by-products of photosynthesis are converted to hydrogen. Chemist David Tiede at the Argonne National Laboratory is working to manipulate the algae genetically. He wants to make the hydrogenase enzyme a more efficient producer of hydrogen gas.

hydrogenase

An enzyme that produces trace amounts of hydrogen during photosynthesis.

In its normal state, the algae manufactures hydrogen gas at an efficiency level of 0.1 percent. That means that only one-tenth of 1 percent of the sunlight absorbed by the algae produces hydrogen. However, Tiede and his team believe that through genetic manipulation, the algae can be pushed to an efficiency level of 10 to 15 percent. According to Tiede, this could produce much more energy than corn ethanol. Tiede says: "We believe there is a fundamental advantage in looking at the production of hydrogen by photosynthesis as a renewable fuel. Right now, ethanol is being produced from corn, but generating ethanol from corn is a . . . much more inefficient process."[35] However, Tiede cautions, utilizing algae for hydrogen production will take a long time: "Hydrogen is one generation or two generations away as the basis for our energy, but we have to start now to find efficient ways to extract it."[36]

Algae-powered fuel cells would rely on a plant that uses photosynthesis to produce hydrogen directly. However, other microorganisms do not need the sun to produce hydrogen. Instead, they use a biological process to separate, or crack, hydrogen from water. At the Massachusetts Institute of Technology, a team led by Angela Belcher, a materials science and biological engineering professor, has genetically modified a harmless virus called M13 to do this work. M13 has been altered to attract and bind with molecules in a catalyst. The catalyst acts as an antenna that captures sunlight. The energy is transferred down the length of the virus like a wire. This energy can line up the molecules in water and split the oxygen from the hydrogen. Belcher expects to have a prototype device that can carry out the process of cracking hydrogen from water by 2012.

This might someday allow virus-based hydrogen pumps to create fuel from water at filling stations.

Microbial Fuel Cells

Microbes can be engineered to produce hydrogen to run standard fuel cells. But some believe that certain bacterial agents can be used as fuel cells themselves. These microbe-powered batteries are called microbial fuel cells (MFCs). Rather than turning food into hydrogen for a fuel cell, MFCs convert biomass directly into electricity.

The concept of MFCs is based on a bacterial microbe called *Geobacter*. The bacteria first came to researchers' attention in 1987, when it was discovered that it could eat highly toxic, oil-based pollutants and radioactive material. *Geobacter* thrives on the toxins and converts them to relatively harmless CO_2. Since the bacteria was discovered, it has been used to clean up hazardous waste sites and oil spills. *Geobacter*, the first known organism that can eat hazardous waste, also consumes organic substances.

Geobacter works by creating tiny, hairlike appendages called pili when consuming food. This process creates a tiny electrical charge, leading scientists to theorize that *Geobacter* might work as an organic fuel cell. The microbial battery would be fed human or animal waste. The electricity generated during the process could be harnessed for power. Professor Derek Lovley and a team of researchers at the University of Massachusetts at Amherst have genetically engineered a strain of *Geobacter* that is eight times more efficient at producing electricity than other strains. In 2010 the researchers were working to create *Geobacter*-based fuel cells that could generate cheap, clean electricity from waste products.

The Bloom Box Boom

Microbial fuel cells only work in theory, and biohydrogen is in the development stages. But a new type of fuel cell, called the Bloom Box, is already being used by some of the most forward-thinking companies in the world.

The Bloom Box was invented by former NASA rocket scientist K.R. Sridhar. He originally created the device to produce oxygen and water for a human colony on Mars. When that did not happen, Sridhar tinkered with his invention so it could power homes and businesses on Earth. He believes the Bloom Box is superior to PEM fuel cells because it does not use expensive platinum and it can burn fuels such as natural gas instead of hydrogen.

The Bloom Box uses a type of battery called a solid oxide fuel cell. It consists of a square ceramic disk about the size of a CD case. The disk is baked from one of the most abundant substances on earth, sand. Each ceramic disk is painted with proprietary inks, green on one side, black on the other. (The chemical basis of the ink is a trade secret.) The inked disks produce electricity when heated to 1,832°F (1,000°C) by natural gas, biogas, ethanol, or hydrogen.

A new and promising type of fuel cell, the Bloom Box, is being used by a select group of companies including eBay. Bloom Energy cofounder K.R. Sridhar shows off his invention outside of the eBay complex in San Jose, California.

A single solid oxide fuel cell disk can light a 25-watt lightbulb. But the disks can be stacked, each one sandwiched between an inexpensive metal plate. A stack of cells the size of a brick can power an average European home. Sixty-four disks provide enough electricity for a coffee shop. When these cells are combined into the 100-kilowatt Bloom Box, which is about the size of 2 large refrigerators, they can produce enough power for 100 average American homes or a small office building.

In 2010 about 20 major corporations in California, including FedEx, Staples, Walmart, eBay, and Google, were using Bloom Boxes to power their operations. Each server costs about $800,000. But 20 percent of the cost was subsidized by the state, and there was also a 30 percent federal tax break for green technology. This allowed the companies to pay about $400,000 for each Bloom Box. At that price, the system pays for itself in 3 years, based on California's commercial electricity cost of 14 cents per kilowatt-hour.

The 5 Bloom Boxes at the eBay campus in San Jose save the company over $11,000 in energy costs each month. EBay's Bloom Boxes are fueled with biogas made from landfill waste, so they are carbon neutral. The energy servers generate more power than the massive array of 3,246 solar panels on the roof of the company's headquarters. Amy Cole, director of eBay's Green Team, explained why the Bloom Boxes are superior: "The solar panels [take] 55,000 square feet and at their peak performance [take] 18 percent of our electricity use off the grid, but that's not at night or on days when we have rain. Running the two side by side over the course of a year, we will get 5 times as much energy from the Bloom system."[37]

One in Every Home

Bloom Boxes generate electricity at a 50 to 55 percent efficiency rate. In comparison, eBay's solar panels produce power at about 12 percent efficiency. However, some Bloom Boxes, like those at Google, are fueled with natural gas and produce CO_2 as a by-product. But the CO_2 emissions from Bloom Boxes compare favorably with standard power generators. A coal plant emits about 2.5 times more emissions than a Bloom Box generating the same amount of electricity. A natural gas plant generates about 1.6 times more CO_2.

A Fuel Cell with Pedals

In 2009 the Tokyo-based Iwatani Corporation developed a hydrogen-powered bicycle. The hydrogen is stored in a refillable cartridge that snaps into place to provide fuel to the PEM fuel cell. The bicycle can be pedaled or run on an electric motor powered by a lithium-ion battery, the type used in hybrid cars. When the battery charge runs low, the fuel cell turns on to recharge the battery while it is in use. The bike is being field-tested by personnel at the Kansai International Airport, where a hydrogen station has been set up to recharge the PEM cartridges.

In their first two years of operation, Bloom Boxes created more than 11 million kilowatts of electricity while cutting 14 million pounds (6.35 million kg) of CO_2 emissions. But Bloom Boxes remain experimental because of a major drawback. Solid oxide fuel cells require extremely high temperatures to function. The individual disks inside burn up and need to be replaced. As a result, each of the thousands of disks in a Bloom Box will need to be replaced twice during the device's 10-year lifespan. That has not stopped Sridhar from envisioning the day when every home will be powered with a scaled-down version of a Bloom Box that costs about $3,000. As Sridhar told *60 Minutes* in a 2010 interview, "In five to ten years, we would like to be in every home."[38] The inventor also believes the boxes will someday provide power to remote villages in Africa and his native India.

Cheaper and More Efficient

By inventing a cheaper fuel cell that runs on readily available natural gas, Sridhar solved major problems associated with PEM cells. Meanwhile, other researchers are investigating methods for low-cost, environmentally friendly hydrogen production. While not yet cost competitive with fossil fuels, experts predict that prices of fuel cell systems will continue to fall as more people purchase the technology. And scientists and researchers are working every day to devise cheaper and more efficient tools for generating renewable energy.

Every year, fuel cells, wind, water, sun, and biomass produce an increasing percentage of the world's energy. While society has taken only the first steps on the renewable energy highway, in the future it is likely that most energy will be from renewable sources. And the smoke, chemical pollution, and greenhouse gases emanating from coal, oil, and natural gas will be a thing of the past.

Source Notes

Introduction: Investing in the Future

1. Quoted in Patience Wheatcroft, "The Next Crisis: Prepare for Peak Oil," *Wall Street Journal*, February 11, 2010. http://online.wsj.com.

2. Quoted in Tiffany Hsu, "Chevron Is Testing Solar's Power," *Los Angeles Times*, March 22, 2010, p. A15.

3. Quoted in White House, "Energy and Environment," March 19, 2009. www.whitehouse.gov.

Chapter One: What Is Renewable Energy?

4. Quoted in Todd Woody, "I.B.M. Opens Energy Lab in Beijing," *New York Times*, March 8, 2010. http://greeninc.blogs.nytimes.com.

5. Quoted in Joseph M. Shuster, *Beyond Fossil Fools*. Edina, MN: Beaver's Pond, 2008, p. 140.

6. Quoted in Lisa Pham, "Waves Start to Make Ripples in Renewable Energy World," *New York Times*, October 20, 2009. www.nytimes.com.

7. Joe Biden, "Progress Report: The Transformation to a Clean Energy Economy," White House, December 15, 2009. www.whitehouse.gov.

8. Quoted in Shuster, *Beyond Fossil Fools*, p. 260.

Chapter Two: Heat and Light

9. Quoted in Peter Johnson, "US Department of Energy Awards UND Researchers $3.5 Million," University of North Dakota, November 11, 2009. www2.und.edu.

10. State Energy Conservation Office, "Texas Geothermal Energy," 2008. www.seco.cpa.state.tx.us.

11. Quoted in David Biello, "Deep Geothermal: The Untapped Energy Source," *Environment 360*, October 23, 2008. http://e360.yale.edu.

12. Quoted in Biello, "Deep Geothermal."

13. Potter Drilling, "Technology Explained," 2010. www.potterdrilling.com.

14. Quoted in Todd Woody, "Major California Solar Project Moves Ahead," *New York Times*, March 17, 2010. http://greeninc.blogs.ny times.com.

15. Quoted in Dave Eyvazzadeh, "Solar Airplane Debuts Before 'Round-the-World' Flight," *Wired*, June 26, 2009. www.wired.com.

Chapter Three: Wind and Water

16. Quoted in Graham Keeley, "Spain's Wind Turbines Supply Half of the National Power Grid," *Times Online*, November 10, 2009. http://business.timesonline.co.uk.

17. Quoted in Alexis Madrigal, "High-Altitude Wind Machines Could Power New York City," *Wired*, June 15, 2009. www.wired.com.

18. David Watson, "Wind Turbines and the Energy in Wind," FT Exploring, 2005. www.ftexploring.com.

19. Quoted in Alexis Madrigal, "Nation's First 'Underwater Wind Turbine' Installed in Old Man River," *Wired*, December 22, 2008. www.wired.com.

20. Quoted in Federal Energy Regulatory Commission, "FERC Approves First Hydrokinetic Device for Existing Hydroelectric Project," December 15, 2008. www.ferc.gov.

21. Quoted in Alexis Madrigal, "Tapping the Vortex for Green Energy," *Wired*, October 28, 2008. www.wired.com.

22. Quoted in Madrigal, "Tapping the Vortex for Green Energy."

23. Quoted in Carl Hoffman, "The Mad Genius from the Bottom of the Sea," *Wired*, June 2005. www.wired.com.

Chapter Four: A New Generation of Biofuels

24. Quoted in Mark Svenvold, "The Biofuels Race," *New York Times*, December 9, 2007. www.nytimes.com.

25. Quoted in Oak Ridge National Laboratory, "Biofuels from Switchgrass: Greener Energy Pastures," 2009. http://bioenergy.ornl.gov.

26. Quoted in Jenny Mandel, "Genetically Modified Tobacco Could Smoke Other Crops as Energy Source," *Scientific American*, January 5, 2010. www.scientificamerican.com.

27. Quoted in Sebastian Blanco, "DOE Announces $786 Million for Third-Generation Biofuels," AutoBlog, May 5, 2009. http://green. autoblog.com.

28. Quoted in *Popular Mechanics*, "10 Big Questions for Maverick Geneticist J. Craig Venter on America's Energy Future," 2009. www. popularmechanics.com.

29. Quoted in Juha-Pekka Tikka, "Craig Venter Has Algae Biofuel in Synthetic Genomics' Pipeline," Xconomy, June 4, 2009. www. xconomy.com.

30. Chris Ayers, "Scientists Find Bugs That Eat Waste and Excrete Petrol," *London Times*, June 14, 2008. www.timesonline.co.uk.

31. Ayers, "Scientists Find Bugs That Eat Waste and Excrete Petrol."

32. Amol Deshpande, "Investing in the Biomass Industry," *BioCycle*, September 2009. www.jgpress.com.

Chapter Five: Hydrogen and Fuel Cells

33. Quoted in Doug Main, "Scientists Seek to Solve Hydrogen Storage Problems," *Washington University Record*, December 2, 2005. http:// record.wustl.edu.

34. Laura Sky Brown, "2015 Toyota Hydrogen Car Will Cost $50,000," Inside Line, May 7, 2010. www.insideline.com.

35. Quoted in Kyle Sherer, "Researchers Investigate Hydrogen-Producing Algae Farms," Gizmag, April 3, 2008. www.gizmag.com.

36. Quoted in Jon Van, "Designer Plants May Produce Hydrogen for Fuel," *Chicago Tribune*, April, 28, 2008. http://articles.chicago tribune.com.

37. Quoted in Ariel Schwartz, "EBay Opens Up About Installing Bloom Boxes and Their Room for Improvement," Fast Company, February 23, 2010. www.fastcompany.com.

38. Quoted in *60 Minutes,* "The Bloom Box: An Energy Breakthrough?" CBS News, February 18, 2010. www.cbsnews.com.

Facts About Renewable Energy

Solar Energy
- According to the Union of Concerned Scientists, there is as much energy in 20 days of sunshine as in all of the world's reserves of coal, oil, and natural gas.
- On days when the sky is mildly overcast, solar systems produce about half as much power as under the full sun, and on very overcast days, as little as 5 to 10 percent.
- According to the U.S. Department of Energy, solar hot water systems reduce the energy needed to produce hot water by 80 percent.
- The price of home solar panels dropped 40 percent between 2008 and 2009 due to an increase in the production of polysilicon, a crucial ingredient used to manufacture the panels.
- In 2010 Toyota began selling the Prius hybrid with an optional rooftop solar panel that powers the ventilation system to keep the car interior cool.

Geothermal Energy
- According to the U.S. Department of Energy, geothermal fields produce only one-sixth of the carbon dioxide of a relatively clean natural gas power plant.
- According to the Geothermal Education Office, the land area required for a geothermal power plant is smaller per megawatt of power generated than for any other type of power plant.
- The United States produces 2,700 megawatts of electricity from geothermal energy, an amount comparable to burning 60 million barrels of oil each year, according to the Geothermal Education Office.

Wind Power
- There are two basic designs of wind electric turbines: vertical axis, called "egg-beaters," and horizontal axis, which are propeller-style machines.

- In 2009 the world's largest wind turbine, installed in Emden, Germany, had a rotor span of 413 feet (126m) and generated 6 megawatts of electricity, enough for about 5,000 European households.
- According to the German Federal Transport Ministry, Germany is planning to install 12,000 megawatts of offshore wind electricity by 2030, the equivalent of 12 medium-size nuclear plants.

Water and Tidal Power

- In 2010, traditional hydropower dams generated about 6 percent of the electricity in the United States, making them the nation's largest source of renewable energy.
- According to the U.S. Department of Interior, the total wave energy potential off U.S. coastlines is equal to the amount of power currently supplied by all of the nation's hydropower dams.
- Ocean engineer Michael Bernitsas says that if 0.1 percent of the energy in the ocean could be harnessed it could support the energy needs of 15 billion people, more than double the earth's current population.
- The world's first commercial wave farm opened in 2008 at the Aguçadora Wave Park in Portugal.
- In 2009, according to the Hydropower Reform Coalition, 55 proposed river hydropower projects were being developed on the Mississippi River between St. Louis and New Orleans. Twenty-two such sites were proposed on the Ohio River, and 27 on the Missouri River.

Biomass Fuels

- About 6 percent of U.S. energy is renewable. Of that renewable energy, 92 percent comes from biomass and hydroelectric power.
- A commercial ethanol plant produces the equivalent of 100 million gallons (378.5 million L) of petroleum every year, the amount of fuel burned worldwide every 41 minutes in 2010.
- Since 2002 the United States has produced more energy from corn ethanol biofuels than from all of its 75,000 hydroelectric dams combined.
- According to the Renewable Fuels Association, there were 170 ethanol distilleries in operation in the United States in 2009 producing 9 billion gallons (34 billion L) of biofuels annually.
- The U.S. Department of Energy estimates that the nation's forest and agricultural lands can sustainably produce 1.3 billion tons (1.18 billion

metric tons) of biomass each year that could be converted into 100 billion gallons (378.5 billion L) of ethanol.

- Researchers at Argonne National Laboratory determined that advances in cellulosic ethanol production could slash the consumption of fossil fuels by 70 percent.

Hydrogen Fuel Cells

- In 2008 there were more than 200 fuel cell vehicles on the road, 175 of them in California.
- According to Larry Burns, vice president of General Motors Research and Development, a network of 12,000 hydrogen stations in the United States would cost $24 billion and put 70 percent of the population within 2 miles (3.2km) of a fueling station.
- In 2008 aerospace company Boeing helped found the Algal Biomass Organization to develop methods for using algae to produce hydrogen jet fuel.

American Solar Energy Society (ASES)

2400 Central Ave., Suite A
Boulder, CO 80301
phone: (303) 443-3130
fax: (303) 443-3212
e-mail: ases@ases.org
Web site: www.ases.org

The mission of the ASES is to inspire energy innovation and speed the transition to a sustainable energy economy. The ASES leads national efforts to increase the use of solar energy, energy efficiency, and other sustainable technologies in the United States. The group publishes *Solar Today* magazine, and leads the ASES National Solar Tour, the largest grassroots solar event in the world.

American Wind Energy Association (AWEA)

1501 M St. NW, Suite 1000
Washington, DC 20005
phone: (202) 383-2500
fax: (202) 383-2505
e-mail: windmail@awea.org
Web site: www.awea.org

The AWEA is a national trade association that represents the wind industry—one of the world's fastest-growing energy industries. The AWEA promotes wind energy as a clean source of electricity for consumers around the world. Publications include the newsletters *Wind Energy Weekly*, *Windletter*, and technology information about wind energy.

California Fuel Cell Partnership (CaFCP)

3300 Industrial Blvd., Suite 1000
West Sacramento, CA 95691

phone: (916) 371-2870
e-mail: info@cafcp.org
Web site: www.cafcp.org

The CaFCP is committed to promoting fuel cell vehicles as a means of moving toward a sustainable energy future, increasing energy efficiency, and reducing or eliminating air pollution and greenhouse gas emissions. The group publishes press releases, newsletters, and information sheets concerning hydrogen power and fuel cell technology.

Idaho National Laboratory (INL)

2525 Fremont Ave.
Idaho Falls, ID 83415
phone: (866) 495-7440
Web site: www.inl.gov

The INL is a science-based, applied engineering national laboratory dedicated to supporting the U.S. Department of Energy's missions in nuclear and energy research, science, and national defense. The INL is the nation's leading research institution in geothermal energy, biomass, wind power, and hydrogen development.

National Biodiesel Board (NBB)

3337A Emerald Lane
PO Box 104898
Jefferson City, MO 65110
phone: (573) 635-3893
fax: (573) 635-7913
e-mail: info@biodiesel.org
Web site: www.biodiesel.org

The NBB represents the biodiesel industry and biodiesel research and development. The NBB's membership is made up of state, national, and international feedstock producers, biodiesel suppliers, fuel marketers and distributors, and technology providers. The NBB publishes *Biodiesel* magazine and the *Biodiesel Bulletin*, a monthly newsletter.

National Fuel Cell Research Center (NFCRC)

University of California
Irvine, CA 92697-3550

phone: (949) 824-1999
fax: (949) 824-7423

The NFCRC was founded to provide leadership in the preparation of educational materials and programs throughout the country. The NFCRC engages undergraduate and graduate students from all disciplines of engineering and the physical and biological sciences and collaborates on courses and team projects.

National Renewable Energy Laboratory (NREL)

1617 Cole Blvd.
Golden, CO 80401-3393
phone: (303) 275-3000
Web site: www.nrel.gov

The NREL is the U.S. Department of Energy's laboratory for renewable energy research, development, and deployment. The laboratory's mission is to develop renewable energy and energy efficiency technologies, advance related science and engineering, and transfer knowledge and innovations to address the nation's energy and environmental goals.

Office of Energy Efficiency and Renewable Energy (EERE)

1000 Independence Ave. SW
Washington, DC 20585
phone: (877) 337-3463
fax: (202) 586-4403
Web site: www1.eere.energy.gov

The EERE invests in clean energy technologies that support research and development of renewable energy technologies, including biomass, geothermal, hydrogen, fuel cells and infrastructure, solar, and wind and water power. Dozens of publications about renewable energy can be downloaded from the EERE Publication Library online.

Renewable Energy Policy Project (REPP)

1612 K St. NW, Suite 202
Washington, DC 20006
phone: (202) 293-2898

fax: (202) 298-5857
e-mail: info2@REPP.org
Web site: www.repp.org

The REPP provides information about hydrogen, solar, biomass, wind, hydro, and other forms of green energy. The goal of the group is to accelerate the use of renewable energy by providing credible facts, policy analysis, and innovative strategies concerning renewables.

For Further Research

Books

Jack Erjavec, *Hybrid, Electric, and Fuel-Cell Vehicles.* Clifton Park, NJ: Thomson Delmar Learning, 2007.

Rex A. Ewing, *Hydrogen—Hot Stuff, Cool Science: Discover the Future of Energy.* Masonville, CO: PixyJack, 2007.

Paul Greenland, *Career Opportunities in Conservation and the Environment.* New York: Checkmark, 2008.

Stuart A. Kallen, *Hydrogen Power.* San Diego: ReferencePoint, 2010.

Krishnan Rajeshwar, Robert McConnell, and Stuart Licht, *Solar Hydrogen Generation: Toward a Renewable Energy Future.* New York: Springer, 2007.

Harriet Rohmer, *Heroes of the Environment: True Stories of People Who Are Helping to Protect Our Planet.* San Francisco: Chronicle, 2009.

Tamara L. Roleff, *Genetic Engineering.* San Diego: ReferencePoint, 2008.

John Tabak, *Solar and Geothermal Energy.* New York: Facts On File, 2009.

Isabel Thomas, *The Pros and Cons of Solar Power.* New York : Rosen Central, 2008.

Sally M. Walker, *We Are the Weather Makers: The History of Climate Change.* Somerville, MA: Candlewick, 2009.

Web Sites

Biomass (www.biomassmagazine.com). An online magazine with the latest news concerning biomass energy–production projects all over the world. The site features biomass blogs, podcasts, and videos and hosts links to *Biodiesel* magazine and *Ethanol Producer* magazine.

California Fuel Cell Partnership (www.fuelcellpartnership.org). This site promotes fuel cell vehicles as a way to reach California's goals for cleaner air and reduced greenhouse gases. Members, including automakers and oil companies, believe fuel cell vehicles must be comparable to or better than the vehicles driven today. Pages explain the benefits of hydrogen power and the technology behind fuel cells, and list events, promotions, and displays concerning fuel cell vehicles.

Energy Efficiency and Renewable Energy (www.eere.energy.gov). A Web site sponsored by the U.S. Department of Energy focused on solar, wind, geothermal biomass, fuel cells, and other renewable energy technologies. The site features comprehensive information about each technology, along with financial opportunities available to green-technology businesses.

Google.org Enhanced Geothermal Systems (www.google.org/egs). Google.org, the philanthropic arm of the Internet search engine giant, is investing millions in geothermal resources to power its electricity-hungry business. This Web site describes various projects being explored and explains EGS technology with a 3-D model and animation of the Cooper Basin EGS project.

On Being a Scientist: A Guide to Responsible Conduct in Research (www.nap.edu/openbook.php?record_id=12192&page=R1). This is a free, downloadable book from the National Academy of Sciences Committee on Science, Engineering, and Public Policy. The 2009 edition provides a clear explanation of the responsible conduct of scientific research. Chapters on treatment of data, mistakes and negligence, the scientist's role in society, and other topics offer invaluable insight for student researchers.

Index

Note: Boldface page numbers refer to illustrations.

Africa, 38, 40
air-conditioning, free, 51
aircraft
  biofuel-powered, 56
  hydrogen fuel cell–powered, 66
  solar-powered, 39
algae, 56, 58–59, 70
Alpha Ventus (offshore wind farm), 44, **45**
American Recovery and Reinvestment Act (2009), 13
Andraka, Chuck, 38–39
Andrianov, Vyacheslav, 57
Archer, Cristina, 42, 43
Argonne National Laboratory, 34
Atienza, Luis, 41
Australia, 31–32
Ayers, Chris, 62

bacteria, 71
batteries, PV cell–charged, 39
Belcher, Angela, 70
Bell Laboratories, 17
Bernitsas, Michael, 49–50, 80
Betz, Albert, 16
bicycle, hydrogen-powered, 74
Biden, Joe, 20, 22
binary cycle power plants, 28–29, 30
biodiesel (B100), 25

biofuels
  cellulosic, 54–56
  as form of solar energy, 64
  first-generation production, 24–25, 53–54, 54, 56, 63
  second-generation production, 54–56
  third-generation production, 56, 57–59
  fourth-generation production, 59–63, 63
biogas, 23, 73
biohydrogen, 69–71
biomass
  described, 23, 53
  microbial fuel cells and, 71
biomass cogeneration, 62–63, **63**
bioreactor landfills, 23–24
biosphere, described, 53
Biotechnology Foundation Laboratories (Thomas Jefferson University), 57–58
Bloom Box fuel cells, 71–74, **72**
Bockris, John O'Mara, 65
Borisjuk, Nikolai, 57
Bradford, Travis, 17
Branson, Richard, 56
British Petroleum, 34–35
Brown, Laura Sky, 68
Bush, George W., 26

carbon dioxide
  Bloom Box and, 73

fourth-generation biofuels and, 59–63

generated by U.S., 12

hydrogen production and, 67

Mycoplasma laboratorium and, 60

second-generation biofuels and, 55

carbon-negative fuels, 59–63

catalysts, 66, 67, 70

Celera Genomics, 61

cellulose, 55

Chevron Corporation, 12–13, 34–35

China
coal use in, 12
investment in clean energy in, 13
petroleum demand in, 11
smart electrical grid in, 14
solar hot water heaters in, 40
Three Gorges Dam, 19
wind power in, 14, 15, 47

Chlamydomonas reinhardtii, 70

Chu, Steven
on DOE, 30
geothermal power and, 29
third-generation biofuels and, 59
wind power and, 41

Clarity FCX fuel vehicle, 67, 68, **69**

Cleantech Innovation Center, 43

coal, 12, 16

coconuts, 56

cogeneration plants, 62–63

Cole, Amy, 73

Conti, Piero Ginori, 22

Cooper Basin (Australia), 31–32

Coproduced, Geopressured, and Low Temperature Projects, 30

coproduction research, 30–31

corn, **54**
disadvantages of use as ethanol, 24–25, 53
energy required to produce ethanol from, 56

Craven, John Piña, 50–51

Cummins, Emily, 38

Deepwater Offshore Wind Consortium (DeepCwind), 44

Department of Energy (DOE)
charge to apply research, 30
grants for biofuel research, 59
grants for geothermal research, 29–30
hydrokinetic energy project, 50
wind power projects, 41, 44

Deshpande, Amol, 64

Diesel, Rudolph, 25

diesel engine cars, 25

Dilley, Philip, 12

DNA, 60

earthquakes from geothermal power plants, 22–23

eBay, 73

eco-fridges, 38

Einstein, Albert, 17

electricity
from Bloom Boxes, 73–74
costs, 16, 36, 42–43
demand in U.S., 41
from geothermal energy, 20, 28, 32
from hydroelectric power, 19

for hydrogen fuel cell
    production, 65
from microbial fuel cells, 71
from ocean thermal energy
    conversion, 51
smart grids, 14
from solar energy, 17, 28, 39
from wind power, 11, 15, 41,
    45, 46
See also specific projects
electrolysis, 26, 65
Energy and Utilities Solutions Lab
    (IBM), 14
enhanced geothermal systems
    (EGS), 31–34
Enviro Energies, 47
environmental concerns
    dams, 19
    earthquakes caused by
        geothermal power plants,
        22–23
    hydrogen fuel cells and
        greenhouse gases, 65, 68
    second-generation biofuels,
        55–56
    solar thermal energy, 35
    See also carbon dioxide
ethanol (ethyl alcohol)
    disadvantages, 24–25
    production
        energy required for, 56
        increase in, 63
        from thermochemical
            conversion process, 55
        in U.S., 24, 53
    uses of, 24
expressed sequence tags, 61

Faraday, Michael, 15

feedstock, 54
fermentation, 55, 62
first-generation biofuel
    production, 24–25, 53–54, **54,**
    56, 63
First Solar, 36
flying electric generators (FEGs),
    42–43
food crops as fuel, 53–54, 56
fossil fuels
    coal, 12, 16
    See also petroleum
fourth-generation biofuel
    production, 59–63
Fridge Lady, 38
fuel cells
    aircraft and, 66
    Bloom Boxes, 71–74, 72
    microbial, 71
    See also hydrogen fuel cells
fuel cell vehicles (FCVs), 67, 68,
    69

Gammons, Brad, 14
garbage, 23
Gelb, Lev, 66
generators, 15, 42
genetic engineering/modification
    Chlamydomonas reinhardtii, 70
    fourth-generation biofuels,
        59–62
    Geobacter, 71
    techniques, 61
    third-generation biofuels, 57–59
genome shotgun technique, 61
Geobacter, 71
geothermal energy
    drilling costs, 32–34, 33
    in Iceland, 24

potential, 20, 22, 29–31, 32

power plants, 22–23, 28

  binary cycle, 28–29, 30

  electricity produced by, 20

  radioactive isotopes and, 22

Geothermal Energy Program
(Southern Methodist
University), 30–31

Germany

  investment in clean energy in,
13

  wind power in, 14, 44, 45

Geysers (geothermal power plant),
22–23

*Global Trends in Sustainable Energy
Investment* (United Nations
Environment Programme), 10

Google

  Bloom Boxes and, 73

  geothermal energy and, 22, 32

  solar thermal energy and, 34–35

greenhouse gases

  fourth-generation biofuels and,
63

  hydrogen fuel cells and, 65, 68

  Mycoplasma laboratorium and,
60

  second-generation biofuels and,
55

Gulf Coast Green Energy, 31

heliostats, described, 34

Honda, 67, 68, **69**

Hoover Dam, 19, **21**

hot water heaters, 40

Huang, Joseph, 51

hydroelectric power

  hydrokinetic turbines and, 48,
49

in U.S., 16, 19, 21

hydrogenase, 70

hydrogen fuel cells

  advantages of, 25, 65

  disadvantages, 26, 66

  NASA PEM technology, 66

  vehicles, 67–69, 69, 74

Hydrogen Fuel Initiative, 26

hydrogen gas Stirling heat engines,
36–39

hydrogen production

  biological, 69–71

  costs, 68, 69

  traditional processes, 65

Hydro Green Energy, 47–48

hydrokinetic turbines, 47–50, **49**

hydrothermal spallation, 33

Hywind, 46

IBM, 14

India, 40

insects, genetically modified,
60–62

Institute for Genomic Research,
61

ions, defined, 67

Italy, 22

Ivanpah Solar Energy Generating
Station, 34–35

Iwatani Corporation, 74

Japan, 74

Jay Oilfield, 31

jet fuel, 56

jet stream winds, 42, 43, 46

Kansas State University, 59

Khosla, Vinod, 53, 54, 60

kilowatt-hour, described, 16

kinetic energy
  defined, 19
  water power and, 19
  wind power and, 14–15
King, Des, 13
kitegens, 42
Klepper, Bud, 54, 55–56

landfill gas-energy projects, 23
lasers for geothermal drilling, 34
lithium-ion batteries, 74
Lovley, Derek, 71
LS9, 60–62

M13, 70
magnetic levitation (maglev) wind
  turbines, 47
Massachusetts Institute of
  Technology (MIT), 70
McLaughlin, Sandy, 56
methane from landfills, 23–24
microbial fuel cells (MFCs), 71
mining, coal, 12
Mississippi River, 48
Moeller, Philip, 48
mountaintop removal, 12
Mycoplasma laboratorium, 60

Namibia, 38
National Aeronautics and Space
  Administration (NASA), 66
National Energy Laboratory of
  Hawaii, 51
National Geographic (magazine),
  30
National Science Foundation
  grants, 59
National Solar Thermal Test
  Facility (NSTTF), 34–35

natural gas
  as Bloom Box fuel, 72, 73
  to produce electricity, 16
  to produce hydrogen, 26, 67, 68
Navy, U.S., and hydrokinetic
  energy, 50
New York Times (newspaper), 16
New Zealand, 22
Nippon Paper Industries USA, 62
Norway, 46
nuclear power plants, 16, 65

Obama, Barack, 13
O'Brien, Terry, 35
ocean-based wind farms, 44–46
ocean power, 19–20
ocean thermal energy conversion
  (OTEC), 50–51
oil. *See* petroleum
Oil 2.0, 60–62
Osborn, Bruce, 38–39
oscillating water columns, 20
Ottaviano, Michael, 20

Pacific Gas and Electric project,
  36
Pal, Greg, 61, 62
parabolic reflectors, 36–39, **37**
Pei, Z.J., 59
PEM (polymer electrolyte
  membrane/ proton exchange
  membrane), 66–68, 74
Peregrine Energy Corporation,
  62–63
petroleum
  industry and green energy, 13
  offshore drilling platforms, 49
  prices, 12
  production, 11–12

spills' clean-ups, 71
vehicular dependency on, 10
worldwide demand for, 11
photosynthesis, 69–70
photovoltaic (PV)
cells, 36, 39
defined, 17
Piccard, Bertrand, 39
platinum, 66, 68
Potter, Bob, 33
Potter Drilling, 32–33
prototypes, defined, 42
pulp and paper industry, 62–63,
63
Puna Geothermal Venture, 28

radioactive isotopes, defined, 22
Range Fuels, 54, 55–56
Red Eléctrica (Spain), 41
refrigeration from solar thermal
energy, 38, 40
Roberts, Bryan, 42

Sandia National Laboratories,
34–35
Scandinavia, 46
Science (journal), 61
second-generation biofuel
production, 54–56
semiconductors, 17–19, **18**
Shepard, David, 42
Sky WindPower, 42
slash, 62
Small Grant for Exploratory
Research, 59
smart electrical grids, 14
Smith, Hamilton, 60
solar electrolysis, 68
solar energy

about, 17
Chevron experimental farm,
12–13
panels, 11
percentage of world's energy
needs provided by, 17
PV cells
cost of electricity production,
16, 36
efficiency of, 19
on rooftops, 35–36
used with Bloom Boxes, 73
Solar Impulse (aircraft), 39
solar thermal energy, 34–35
hot water from, 40
refrigeration from, 38, 40
Stirling heat engines, 36–39
Southern California Edison solar
project, 36
Southern Methodist University,
30–31
Spain, 41
spallation, defined, 33
Sridhar, K.R., 71–72, **72**, 74
StatoilHydro, 46
steam-reforming, 65
Stirling, Robert, 36
Stirling heat engines, 36–39
Stover, Mark, 48
sulfur dioxide generated by
U.S., 12
SunCatcher, 37–39
sustainable energy, described,
13
switchgrass, 54, 56
synthetic gas (syngas), 55–56
Synthetic Genomics, 60

Tester, Jefferson, 32

thermochemical conversion
  process, 55
thermodynamics, defined, 14
thin-film PV, 36
third-generation biofuel
  production, 56, 57–59
Thomas Jefferson University,
  57–58
Three Gorges Dam, 19
tidal barrage generators, 19, 20
tidal power, 19–20
Tiede, David, 70
tobacco, 57–58, **58**
20% Wind Energy by 2030, 41

United Nations Environment
  Programme, 10
United States
  biomass cogeneration in,
    62–63, 63
  coal use in, 12
  electricity demand in, 41
  ethanol production in, 24, 53
  See also Department of Energy
    (DOE); specific projects
U.S. Geological Survey (USGS),
  29

vehicles, fuel cell, 67–69, **69**,
  74
Venter, J. Craig, 59–60, 61
Virgin Atlantic, 56
viruses, 70–71
volcanoes, 28
vortex (vortices), defined, 50

Vortex Induced Vibrations
  Aquatic Clean Energy
  (VIVACE) system, 50

waste vegetable oil for diesel
  vehicles, 25
water, use in biofuel production, 56
water power
  hydroelectric
    hydrokinetic turbines and, 48,
      49
    in U.S., 16, 19, 21
  kinetic energy of, 19
  tidal, 19–20
  from vibrations, 48–50, 49
Watson, David, 47
wave power, 20
wellbores, defined, 31
Wells turbines, 20
wind energy
  high-altitude, 42–43, 46
  need for backup generators, 41
  potential U.S. electrical
    production from, 45
wind power plants, **11**
  in China, 14, 15
  cost of electricity production, 16
  efficiency of, 15–17
  maglev, 46–47
  offshore, 44–46
  in Spain, 41
  underwater, 47–48
wood mass, 62–63, **63**

Yuan, Wayne, 59

Picture Credits

Cover: iStockphoto.com

AP Images: 15, 18, 33, 37, 45, 49, 63, 69, 72

iStockphoto.com: 8, 9, 11, 21, 24, 54, 58

About the Author

Stuart A. Kallen has written more than 250 nonfiction books for children and young adults over the past 20 years. His books have covered countless aspects of human history, culture, and science, from the building of the pyramids to the music of the twenty-first century. Some of his recent titles include *How Should the World Respond to Global Warming? National Security, Toxic Waste,* and *Hydrogen Power.* Kallen is also an accomplished singer-songwriter and guitarist in San Diego, California.